Sajad Kiani

Emerging Platforms in Fluid Flow Transport: Surfactants, Polymers, and Nanoparticles

Emerging Platforms in Fluid Flow Transport: Surfactants, Polymers, and Nanoparticles

Sajad Kiani, Shirin Alexander, and Andrew R. Barron

MiDAS Green Innovations
2021

Cover image © Susan Vineyard (2018)

First Printing: 2021

ISBN 978-1-8384167-0-6

MiDAS Green Innovation, Ltd
Swansea, SA1 8RD, UK

www.midasgreeninnovation.com

Dedication

This book is dedicated to our families for all their support.

Contents

Acknowledgements

Thank you to all academic reviewers and contributors who participated in the production of this book, including:

- Professor Randall Seright of the Petroleum Department at New Mexico Institute of Mining and Technology (New Mexico Tech)
- Dr Mohammadali Ahmadi of the Department of Chemical and Petroleum Engineering at University of Calgary
- Dr Goshtasp Cheraghian of the Braunschweig Pavement Engineering Centre (ISBS) at Technische Universität Braunschweig.

Preface

Chemical flooding in oil reservoirs began as far back as the early 1920s. Since the early 1990s, chemical flooding has progressed considerably facilitating huge amounts of oil production. Attention has most recently been turned towards chemicals such as surfactants, polymers, and nanoparticles to enhance fluid flow by overcoming strong chemical bonding between rock-fluid and fluid-fluid interfaces. This book begins with the fundamentals aimed at those with basic knowledge of petroleum engineering. The author's objective is to deliver an uncomplicated introduction to the field of chemical flooding for oil recovery.

Also discussed are chemicals that enhance the effectiveness of each chemical flooding. Chapters have been kept as concise as possible, ensuring to end with a clear summary to allow the reader to grasp the overall information. In the first chapter, a brief introduction to oil recovery background is given before moving on to topics such as fluid flow transport in Chapter 2 and chemical flooding's including surfactant in Chapter 3, polymer in Chapter 4, nanoparticles in Chapter 5, and finally the challenges and emerging technologies in Chapter 6. Furthermore, Chapters 3 to 5 provide a unique insight and discus key elements, not found anywhere else. This book illustrates the reasons chemicals are required and why they should have been studied much earlier in order to maximize oil production. In general, this book should be considered essential reading for those interested in the application of chemical floods using surfactants, polymers, and nanomaterials.

Symbols and Abbreviations

AM	acrylamide
APG	alkyl polyglycerol-side
API	American Petroleum Institute
CBM	coal-bed methane
CFB	circulating fluidized bed
CMC	critical micelle concentration
CNTs	carbon nanotubes
CSG	coal seam gas
CSS	cycle steam simulator
CTAB	cetyltrimethylammonium bromide
DLE	double layer expansion
DPR	disproportionate permeability reduction
EDA	ethylenediamine
EO	ethylene oxide
EO-PO	ethylene oxide-propylene oxide
EOR	enhanced oil recovery
FAME	fatty acid methyl ester
FBA	fluorobenzoic acid
HF	hydraulic fracturing
HLB	hydrophilic-lipophilic balance
HNPs	hairy NPs
HPAI	high-pressure air injection
HPAM	hydrolyzed polyacrylamide
HPG	hyperbranched polyglycerol
HPHT	high pressure high temperature
HTHS	high temperature high salinity
IEA	International Energy Agency
IEP	isoelectric point
IFT	interfacial tension
ILs	ionic liquids
ISC	in-situ combustion
Kr_i	injected relative permeability
Kr_o	oil relative permeability

ODS	oil-oxidative desulfurization
OOIP	original oil-in-place
λ_i	injected fluid mobility
λ_o	oil fluid mobility
LSES	low surface energy surfactant
M	mobility ratio
MIE	multi-component ion exchange
μ_i	injected fluid viscosity
μ_o	oil fluid viscosity
η	viscosity
NC	nanocellulose
NPs	nanoparticles
nZVI	nano zero-valent iron
PEO−PPO	poly(ethylene oxide)−poly(propylene oxide)
PEG	polyethylene glycol
PAAm	polyacrylamide
PAH	polycyclic aromatic hydrocarbon
PAMAM	2-poly(amidoamine)
P_c	capillary pressure
PDMS	polydimethylsiloxane
PG-NPs	polymer grafted nanoparticles
PO	propylene oxide
POE–POP–	polyoxyethylene–polyoxypropylene–
POE	polyoxyethylene
PTSA	pyrene tetrasulfonic acid
PV	pore volume
PVA	polyvinyl alcohol
PVI	pore volume injected
QDs	quantum dots
R-CO$_2$H	carboxylic groups
R-O-[PO$_4$]$_3$-	phosphate
R-O-SO$_3$-	sulfate
R-SO$_3$-	sulfonate
SANS	small-angle neutron scattering
SAGD	steam-assisted gravity drainage
SAGP	steam and gas push
SEM	scanning electron microscopy

SCW	supercritical water
SDS	sodium dodecyl sulfate
SHC	solid heat carrier
SHMP	sodium hexametaphosphate
TVP	thermoviscosifying polymer
VES	viscoelastic surfactants
WAG	water alternate gas
WEO	World Energy Outlook
XPS	X-ray photoelectron spectroscopy

Chapter 1: Introduction

Despite a push by many nations towards green energy production, the global consumption of oil has not abated due to the increased economic demand of developing regions (Figure 1.1). Two general approaches are available to meet the demand for oil production:
- developing more reservoirs,
- increasing the production of already in-place reserves.

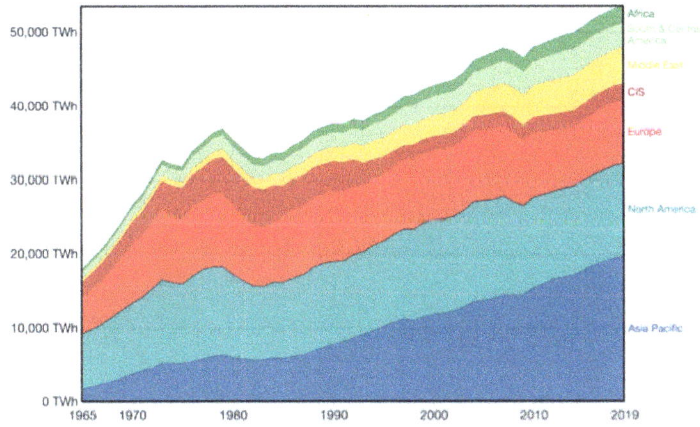

Figure 1.1: Regional annual oil consumption, measured in terawatt-hour (TWh) equivalents. Adapted from BP Statistical Review of Global Energy (2020). Copyright: OurWorldData.org (2020).

The first of these requires increased exploration and production, which was traditionally costly and necessitates accessing new geographic locations, with attendant political and environmental implications. The development of unconventional shale oil has partially offset these issues in the US and Canada by opening up previously untapped reserves. In developing shale oil, the industry has moved from a "difficult to find, easy to produce" to an "easy to find, difficult to produce" scenario. However, in many geographic regions, new reservoir opportunities are limited and if additional oil production is not developed, the concept of peak oil (Simons, 2006) will become a reality, along with the economic, social and political changes that will transpire.

Oil production is generally divided into three phases, which are primary, secondary, and tertiary (Figure 1.2). Oil production using the reservoir's natural energy (natural pressure) or artificial lift (pump, gas lift) via a single well is categorized as *primary oil recovery*. Unfortunately, natural driving forces are insufficient and primary oil recovery does not produce anywhere near the required amount of the original oil-in-place (OOIP) (Muggeridge et al., 2014); values are typically 5-15%. The systems behind primary oil recovery are primarily depletion/gas cap/aquifer drive, gravity drainage, and rock/liquid expansion.

Once oil production declines, the addition of energy (frequently through injection of gas and water into the oil formation) is required to retain or raise oil production levels to extract the remaining fluids and gas in the reserves. This additional energy is frequently via injection of gas and water into the formation rock through single or multiple injection wells. The addition of gas and/or water, or a mixture of both (gas/water) in the *secondary oil recovery* creates a considerably large amount of oil extraction (15-40% of OOIP). Unfortunately, both primary and secondary methods suffer from several limitations, including a rapid decrease in reservoir pressure and insufficient oil recovery due to the unfavorable mobility ratio (the difference between injected fluid and oil displacement on rock surfaces) among fluids, coning problems (water and gas), and lack of adequate sweep efficiency (Ali & Thomas, 1996). To overcome the above-mentioned issues *tertiary enhanced oil recovery* (TEOR) methods are employed, some of which include chemical, thermal, and gas miscible displacement processes (Green & Willhite, 1998). The TEOR phase of progress can increase oil recovery by 50–60% in optimum conditions. However, the techniques discussed to enhance oil recovery are justifiable considering that worldwide recovery averages at 35% leaving the remaining 65% unrecoverable. The various levels of enhanced oil recovery (EOR) are represented schematically in Figure 1.2.

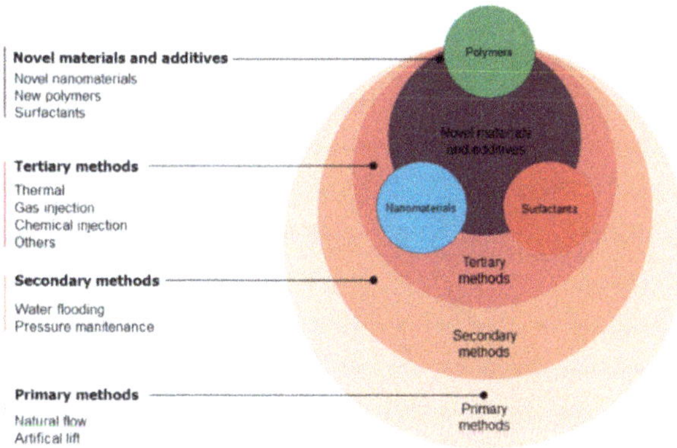

Figure 1.2: Schematic representation of EOR classification and injection into reservoir. Reproduced with permission from R. Al-Mjeni, S. Arora, P. Cherukupalli, J. van Wunnik, J. Edwards, B. J. Felber, O. Gurpinar, G. J. Hirasaki, C. A. Miller, and C. Jackson. Has the time come for EOR, *Oilfield Rev.*, 2010/2011, 22, 16. Copyright: Schlumberger (2010).

The contemporary debate over oil production decline has its roots in the understanding of strong chemical interactions between oil, rock, and other fluids such as connate water in the source rock. The subsidence of hydrocarbons over millions of years elevates their temperature and pressure, launching the process of organic hydrocarbon maturation. This process involves two fundamental sub-steps that firstly converts the mature organic compounds into insoluble supra-molecule kerogen and converts into lighter hydrocarbon fractions as organic maturation upsurges in the long term.

It is to be noted that kerogen will break off into smaller hydrogen-rich substances with a liquid form in the second step. Today, many reservoirs have lost their lighter hydrocarbon fractions due to declining reservoir pressure and the current rate of oil production. The liberation of oil from the host rock typically commences at changing physicochemical interactions between solid-oil-water systems and continues by large oil displacement. The main interactions between the rock-oil-fluid systems have been illustrated in Figure 1.3.

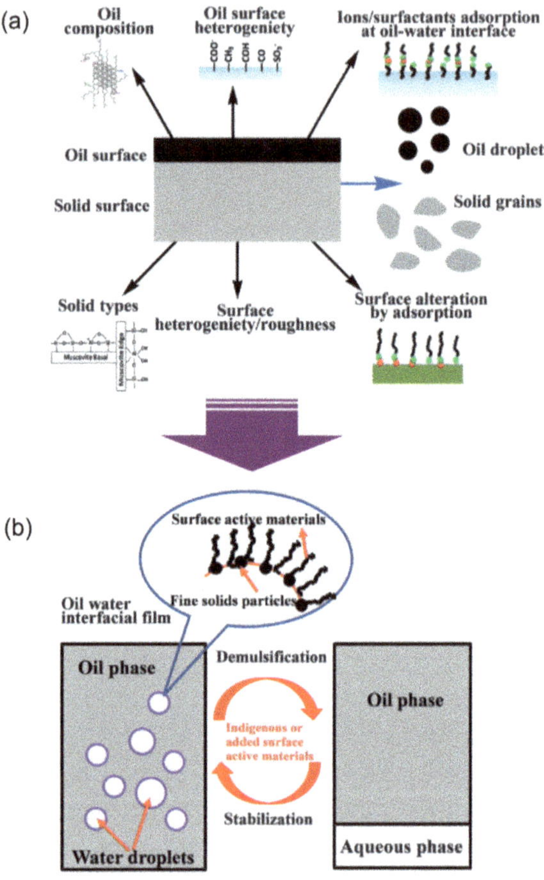

Figure 1.3: Main interactions for oil liberation through porous media with utilization of novel material: (a) liberation of unconventional oils from the surfaces of host solids/sands, which is mainly affected by parameters such as oil composition, oil surface heterogeneity, oil–water interfacial properties, solid surface characteristics and water chemistry; (b) separation of liberated unconventional petroleum from water, involving stabilization and destabilization of oil–water emulsions by chemicals or mineral particles. Adapted from L. He, F. Lin, X. Li, H. Sui, and Z. Xu, Interfacial sciences in unconventional petroleum production: from fundamentals to applications, *Chem. Soc. Rev.*, 2015, 44, 5446. Copyright: Royal Society of Chemistry (2015).

Generally speaking, TEOR approaches can be described by one of two techniques:

- thermal,
- non-thermal.

Thermal methods are essentially employed to reduce oil viscosity and have been utilized in heavy and extra-heavy oil reserves, as well as tar sands (Al-Murayri et al., 2016; Babadagli, 2020). On the other hand, non-thermal methods involve fluid (miscible or immiscible) injections to improve the mobility ratio ≤ 1. A miscible injection is an opposite approach to water flooding secondary recovery, in that instead of injecting water to displace the desired hydrocarbon, injection of a miscible compound (CO_2) or solvent (LPG or alcohols) essentially dissolves the hydrocarbon creating a mixture with increased mobility (Miller & Sorrell, 2014).

Current chemical flooding methods are divided into three main classes:

- surfactant,
- polymer,
- alkaline flooding.

In each case, the aim is to use chemicals to change the physicochemical interactions between the rock-fluid and fluid-fluid interactions (Trabelsi et al., 2011). For example, causing a significant change to the wettability of sandstone rock from oil-wet to water-wet. However, significant technical challenges remain, and it is in this field that nanoparticles (NPs), surfactants, and polymers show great potential (Alameri et al., 2014; Li et al., 2010; Lago et al., 2012).

Currently, the use of EOR methods have led to an additional production of 2.5×10^6 barrels per day, and it is estimated that EOR could facilitate the extraction of as much as 3×10^{12} barrels of trapped oil. Figure 1.4 shows that the proposed future relative oil production, in the USA and the rest of the world, is influenced by EOR methods. For example, in 2024, the EOR market size is expected to increase to over 25% (~\$89 billion) due to higher demand within industries such as automobiles. Furthermore, it is expected that primary recovery will decrease while the amount of oil production employing secondary and tertiary methods will increase. The greater use of tertiary EOR methods in the USA as compared to the rest of the world is a function of the need to enhance recovery from shale reservoirs in the former (Muggeridge et al., 2014).

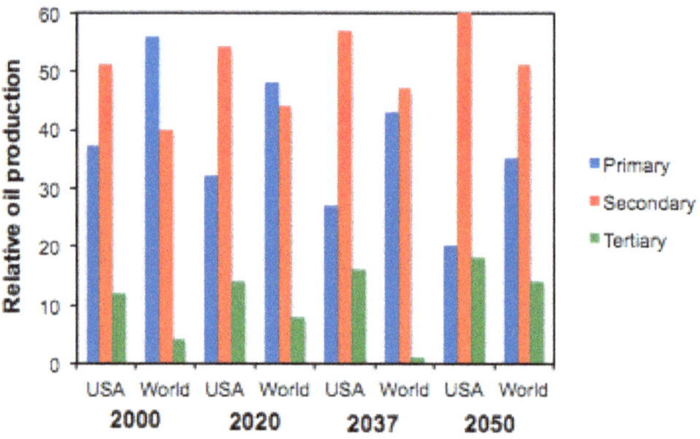

Figure 1.4: Proposed future relative oil production, in the USA and the rest of the world, influenced by EOR methods. Reproduced with permission from G. J. Stosur, EOR: Past, present and what the next 25 years, *SPE J.*, 2003. Copyright: SPE (2003).

Figure 1.5 estimates the oil reserves including light, heavy, and extra-heavy oil across the world. A significant portion of the oil is produced by Saudi Arabia (266.8 billion barrels of oil) from its Ghawar Field (the largest oil reserve in the world), which appear to be running out as a result of the high-pressure drop in the reservoir. Saudi Arabia produces approximately 10 million oil barrels per day this means they can only produce oil for the next 73 years (i.e., their reserves will be completely used by 2090). Running out of oil is shocking news for the world's population while other alternative fuels and technologies such as hydrogen, wind, and solar cell are still being brought on-line at a scale to provide the increasing global energy demands. According to World Energy Outlook (WEO), using EOR programs has led to estimated jumps of 5 million barrel per day by the year 2030. Therefore, the need for affordable technologies using current and new chemicals to extract more than 60% of the remaining oil is imperative.

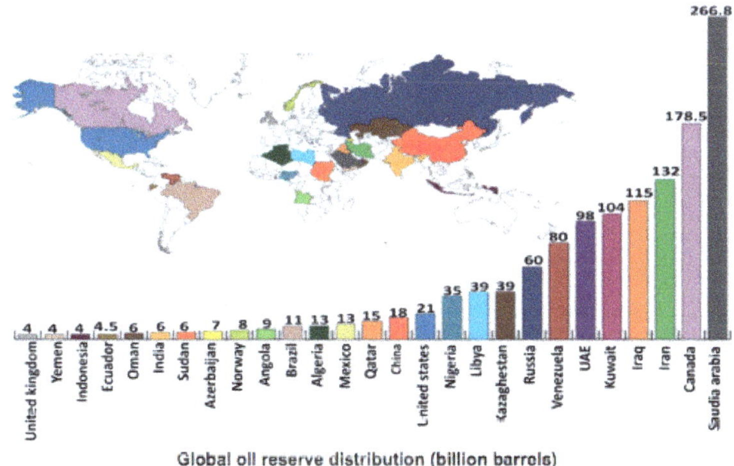

Figure 1.5: Distribution of heavy and light oil reservoirs across worldwide. Copyright: US Energy Information Administration (2008).

The main heavy and recoverable oil reservoirs are located in North America (Figure 1.6). Hydraulic fracturing (HF) is the main recovery technique available to sweep heavy oil out of tight and shale reservoirs. HF technique creates a microstructure within the oil-containing formation and leads to a marked improvement in the permeability. However, it should be noted the oil recovery factor by HF is currently ~8%, the main reason behind this figure is the lack of rock permeability (0.001 to 0.0001 mD) in shale and tight reserves. The rock permeability in shale reservoirs is 1000 times less than the value of conventional reservoir rock (Thomas et al., 2008).

Figure 1.7 illustrates predicted global all-liquids production by 2035 across the world according to the International Energy Agency (IEA) report. The IEA reports anticipate a decrease in crude oil production per day of 42.4 million barrels from 2011 to 2035 in existing oilfields. However, it expects a slight decrease in total oil production due to adding and developing further undiscovered oil reserves. In general, IEA suggested no remarkable decline in oil production before 2035.

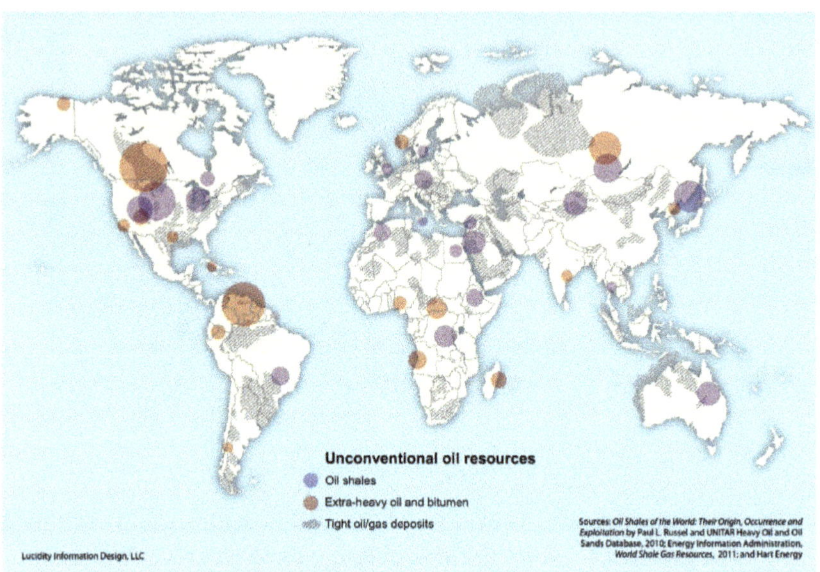

Figure 1.6: Distribution of tight, shale, and heavy oil reservoirs depicts the main highly viscous oil resources are currently in the USA, Canada, Venezuela, and Russia (US Energy Information Administration, 2011).

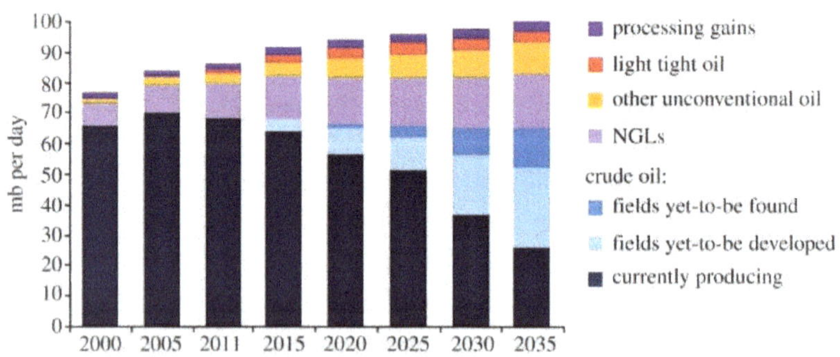

Figure 1.7: Predicted global petroleum production by 2035. Reproduced with permission from R. G. Miller and S. R. Sorrell, The future of oil supply, *Philos. Trans. A Math. Phys. Eng. Sci.*, 2014, 372, 20130301. Copyright: The Royal Society (2013).

Laboratory studies, as well as pilot experiments, have shown that polymer, surfactant, and nanoparticles are the most effective chemicals in the reservoirs for maximum oil recovery. Furthermore, some advantages, such as minimal scaling and emulsification, have recently come to fruition in simulated reservoirs. This demonstrates the potential for ground-breaking results within oil recovery via these synthetic novel chemicals (Ali & Thomas, 1996; Lago et al., 2012).

Bibliography

S. M. Ali and S. Thomas, The promise and problems of enhanced oil recovery methods, *J. Can. Pet. Tech.*, 1996, **35**, PETSOC-96-07-07.

BP Statistical Review of Global Energy, BP (2020).

R. Al-Mjeni, S. Arora, P. Cherukupalli, J. van Wunnik, J. Edwards, B. J. Felber, O. Gurpinar, G. J. Hirasaki, C. A. Miller, and C. Jackson, EOR: Past, present and what the next 25 years, *Oilfield Rev.*, 2010/2011, **22**, 16.

D. W. Green and G. P. Willhite, *Enhanced Oil Recovery*, Henry L. Doherty Memorial Fund of AIME, Society of Petroleum Engineers, Richardson, Texas (1998).

L. He, F. Lin, X. Li, H. Sui, and Z. Xu, Interfacial sciences in unconventional petroleum production: from fundamentals to applications, *Chem. Soc. Rev.*, 2015, **44**, 5446.

S. Lago, H. Rodríguez, M. K. Khoshkbarchi, A. Soto, and A. Arce, Enhanced oil recovery using the ionic liquid trihexyl (tetradecyl) phosphonium chloride: phase behaviour and properties, *RSC Adv.*, 2012, **2**, 9392.

A. Muggeridge, A. Cockin, K. Webb, H. Frampton, I. Collins, T. Moulds, and P. Salino, Recovery rates, enhanced oil recovery and technological limits, *Philos. T. R. Soc. A*, 2014, **372**, 20120320.

R. G. Miller and S. R. Sorrell, The future of oil supply, *Philos. Trans. A Math. Phys. Eng. Sci.*, 2014, **372**, 20130301.

G. J. Stosur, EOR: Past, present and what the next 25 years, SPE-84864-MS, SPE J., 2003

M. Simmons, *Twilight in the Desert: The Coming Saudi Oil Shock and the World Economy*, 1st edition, Wiley, New Jersey (2006).

S. Thomas, Enhanced oil recovery-an overview, *Oil Gas Sci. Technol., Rev. IFP*, 2008, **63**, 9

US Energy Information Administration (EIA), 2008.

24

Chapter 2: Overview of Fluid Flow Transport

Main parameters affecting EOR

In general, the main factors affecting successful enhanced oil recovery (EOR) projects are:

- mobility ratio,
- fluid viscosity,
- salinity,
- pH,
- rock porosity,
- permeability,
- rock lithology,
- injected fluid,
- temperature.

The most effective way of measuring reservoir efficiency *(E)* is by determining the fraction of oil that has been recovered from a zone swept by a waterflood or other displacement process, as defined by Equation 2.1, where V_{oi} = volume of oil at start of flood and V_{or} = volume of oil remaining after flood.

$$E = \frac{(V_{oi} - V_{or})}{V_{oi}} \tag{2.1}$$

Recovery efficiency is determined as follows; the microscopic sweep efficiency (displacement efficiency, E_D) and macroscopic sweep efficiency (volumetric efficiency, E_V).[1] Fluid mobilization at pore scale is defined as microscopic fluid displacement contributes to maximum

---

[1] Sweep efficiency is defined as a measure of the effectiveness of an enhanced oil recovery process that depends on the volume of the reservoir contacted by the injected fluid. The volumetric sweep efficiency is an overall result that depends on the injection pattern selected, off-pattern wells, fractures in the reservoir, position of gas-oil and oil/water contacts, reservoir thickness, permeability and areal and vertical heterogeneity, mobility ratio, density difference between the displacing and the displaced fluid, and flow rate.

oil recovery when the mobility ratio between displaced and displacing fluid is less than 1. Also, fluid mobilization at reservoir scale is defined as macroscopic fluid displacement, which is the mobility ratio between injected fluid and oil. The factors determining this depend on the type of EOR implemented. It is important to note that optimal oil sweeping directly relates to miscible or immiscible injection in the reservoirs. For instance, higher water flooding leads to changes in capillary pressure (P_c) and interfacial tension (IFT) between fluids and consequently leads to lower relative wettability in the long term (Buckley & Fan, 2007).

Mobility ratio

The fluid mobility is formed as follows; relative permeability divided by its viscosity. Fluid mobility is determined by properties of subsurface rock formation, fluid viscosity, and permeability. Mobility ratio is calculated by the mobility of injected fluid (e.g., gas or water) divided by the mobility rate of displaced fluid according to Equation 2.2, where λ_i and λ_o represent the mobility of the injected fluid (i = gas or water) and crude oil (o), Kr_i and Kr_o represent the relative permeabilities, and μ_i and μ_o represent the fluid viscosity of the injected fluid and crude oil, respectively.

$$\text{M} = \frac{\lambda_i}{\lambda_o} = \frac{Kr_i/\mu_i}{Kr_o/\mu_o} = \frac{Kr_i \cdot \mu_o}{Kr_o \cdot \mu_i} \qquad (2.2)$$

Crude oil has better movement than water/gas when M is 1 (favorable condition). Note: the relative permeability of a phase is measured by the ratio of the phase to absolute permeability, Equation 2.3.

$$k_{ri} = \frac{k_i}{k} \qquad (2.3)$$

Fluid viscosity

Suitable fluid viscosity is one of the main factors in successful EOR flooding. Unfavorable fluid mixing leads to changes in the value of macroscopic oil sweep efficiency. Therefore, proper control of flooding is associated with preventing viscous fingering (Doorwar & Mohanty, 2017). Viscous fingering occurs at a gas-oil interface when a less viscous fluid such as gas displaces a high viscous fluid like water (Figure 2.1). As a result of this hydrodynamic instability at low Reynolds numbers, viscous fingering will be enhanced. Therefore, viscosity differences between gas-liquid are due to huge amounts of trapped oil in the pore-throat (increasing mobility ratio and residual oil). Moreover, when the water breakthrough occurs by fluid viscous fingering, the water tends to push out more than oil, causing the least resistance in oil pathways (Doorwar & Mohanty, 2017). Fingering effect and gravitational segregation are the main reasons behind poor macroscopic sweep efficiencies. Early gas breakthrough often occurs in gas injection specifically by CO_2 (Chen et al., 2015).

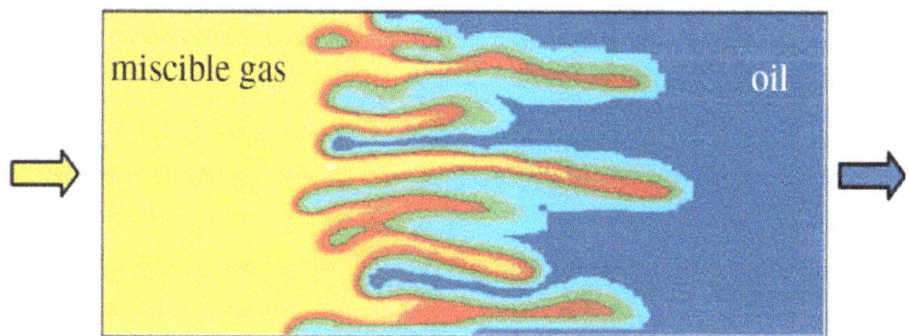

Figure 2.1: A numerical simulation of viscous fingering seen when low-viscosity gas displaces higher viscosity oil. Reproduced with permission from A. Muggeridge, A. Cockin, K. Webb, H. Frampton, I. Collins, T. Moulds, and P. Salino, enhanced oil recovery and technological limits, *Phil. Trans. R. Soc. A,* **2014, 372, 20120320. Copyright: Royal Society (2014).**

Figure 2.2 exemplifies the process of oil trapping based on water-wet systems through porous media. As can be seen, the pores are saturated with oil, after which, water flooding begins (capillary effects) resulting in a thick water film and finally, upon completing these stages, the water films join and oil continuity is lost, i.e., interfacial tension reduction (Buckley & Fan, 2007).

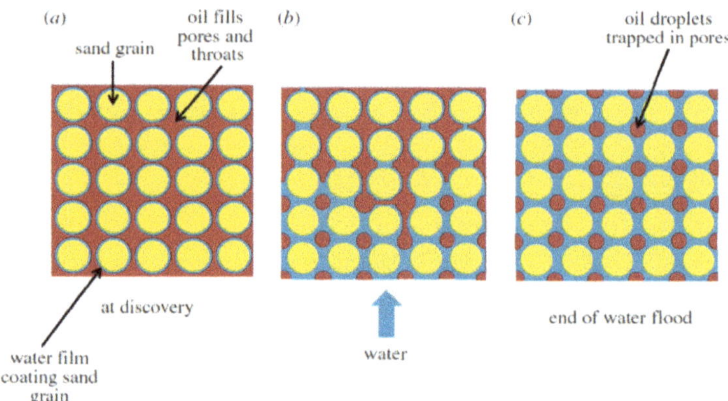

Figure 2.2: Schematic representation of oil trapping process based on water-wet systems through porous media. (a) At discovery, the sand grains are coated with a thin water film and the pores are filled with oil; (b) as water flooding progresses the water films become thickens until (c) the water films merge, and oil continuity ceases. Reproduced with permission from A. Muggeridge, A. Cockin, K. Webb, H. Frampton, I. Collins, T. Moulds, and P. Salino, enhanced oil recovery and technological limits, *Phil. Trans. R. Soc. A*, 2014, 372, 20120320. Copyright: Royal Society (2014).

Groundwater chemistry and rock properties

Groundwater in a subsurface rock formation is one of the most important parameters to transmit fluids in the host rock (Farajzadeh et al., 2019). The rock formation is mostly sedimentary rocks composed of silica, carbonate and clay minerals with a variety of salts. Experimental studies have shown the importance of strong adhesion interactions such as the van der Waals for detachment of oil on the rock formation. It is expected that strong affinities at the oil-rock interface would decrease by manipulating water flooding with optimum brine (low salinity water injection) in the presence and absence of nanoparticle, polymer and surfactant (Ali et al., 2018; Esfandyari Bayat et al., 2014; Myint & Firoozabadi, 2015; Gassara et al., 2015; Nasralla et al., 2013). This flooding leads to significant changes to the chemistry of water and rock formation. They can be used as a promising EOR technique for light to mature hydrocarbon reservoirs where the oil production has declined through physicochemical interactions between solid-liquid and liquid-liquid and improved oil recovery by 5-25%.

Formation water mainly contains monovalent and divalent cations such as Na^+, K^+, Ca^{2+}, Mg^{2+}, Cl^-, SO_4^{2-}, NO_3^-. According to double layer expansion (DLE) and multi-component ion exchange (MIE) theory, these ions have a central role in a successful EOR method (Zhang et al., 2007), because they have a dominant effect on rock wettability alteration and lead to chemical aggregation and consequently pore plugging through the pores (Pouryousefy et al., 2016). The cation exchange capacity on the rock is considered one of the main equilibria between water-oil at reservoir rocks as it is fluid injected and has recognized new equilibria with the formation. Typical divalent cations such as Mg^{2+} and Ca^{2+} in the reservoir can provide strong adsorption after fluid injection on rock surfaces due to multi-component ion exchange (MIE) theory. Multivalent cations at sand surfaces can be bonded to heavy oil polar components and form organometallic complexes, hence the oil-wet condition of the rock surface (Pouryousefy et al., 2016).

The ideal water chemistry design should retain ion repulsive forces across the oil-sand interface because the kinetics of these forces can attribute the increase in adsorption of ions such as Na^+ and Ca^{2+} on sand surfaces. One of the most common surfactants, which have been utilized in water chemistry alteration, is sodium dodecyl sulfate (SDS, Figure 2.3), which can mitigate the adhesion force between fluid-rock interactions and decrease solid hydrophobicity.

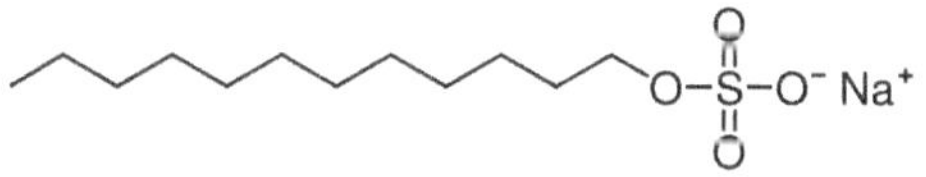

Figure 2.3: The structure of sodium dodecyl sulfate (SDS).

Analysis of wettability alteration and interfacial tension

Wettability alteration has a major role in the location flow control and oil sweep efficiency within porous media. It is strongly depended on geochemistry interactions between oil/water/brine, and geological host solids (clay, sand, rock) (Buckley & Fan, 2007; Moghaddam et al., 2015). The wetting phenomenon may be split into the following categories based upon the measured contact angle:

- hydrophilic (water-wet) = 0-30°,
- hydrophobic (oil-wet) = 150-180°,
- bi-wettable ≈ 90°.

Relative permeability (rel-perm) and core flooding experiments can be used as qualitative methods. Contact angle and interfacial tension (IFT) measurements (adhesive and cohesive forces) are common strategies to fully explore the wetting degree by a force balance on solid-fluid and fluid-fluid interfaces (Table 2.1). IFT is described as the amount of molecule adsorption at the surface by making a contact boundary, which refers to Gibbs-free energy equation. Gibbs equation studies suggested that IFT measurements are based on the pendant drop, spinning drop, or sessile drop in the presence of fluids. These IFT measurements are responsible for enhancing oil liberation from the rock surface (Moeini et al., 2014; Trabelsi et al., 2011). IFT among fluids, salinity, pH, acidic number, asphaltenes resin fractions, viscosity, and HPHT condition are the main factors to change the surface wettability of rock.

Contact angle (°)	Degree of wetting	Strength of solid / liquid interactions	Strength of liquid / liquid interactions
$\theta = 0$	Perfect wetting	Strong	Weak
$0 < \theta < 90$	High wettability	Strong	Strong
$90 < \theta < 180$	Low wettability	Weak	Weak
$\theta = 180$	Perfect non-wetting	Weak	Strong

Table 2.1: Summary of wetting in relation to experimental contact angle.

Spontaneous imbibition method is the most common method for measuring the wettability of a core-plug (Moeini et al., 2014; Tagavi-far et al., 2019). Assessing core plug using the imbibition method depends heavily upon the fluid viscosities, temperature, salinity, relative permeability, and initial core saturation (Figure 2.4). Figure 2.4

shows wetting behavior (the contact angle) of water and oil droplets on the solid substrate.

Through numerous experimental tests, it has been shown that the rock wettability can be modified by a thermal process or chemical additives such as saline injection and cation exchange, surfactants, nanoparticles, and polymers (Esfandyari Bayat et al., 2014). Furthermore, physical properties such as fluid distribution, capillary pressure (P_c), and relative permeability (Kr_o) can be utilized to evaluate wettability alteration. Carboxylic groups ($R\text{-}CO_2H$) in heavy and extra-heavy oils provide one of the most important acid-base interactions on the rock surfaces in carbonate reservoirs. Carboxylic interactions are a standard approach to generating strong adsorptions between rock-fluid.

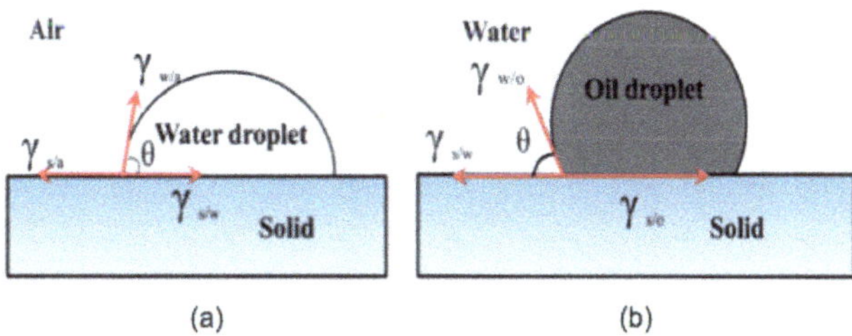

Figure 2.4: Wetting behavior of two systems at the interfaces: (a) solid/water/air, and (b) solid/oil/water system. The contact angle is defined as an angle obtained by the Young equation when a droplet meets a solid substrate. Reproduced with permission from L. He, X. Li, G. Wu, F. Lin, and H. Sui, Distribution of saturates, aromatics, resins, and asphaltenes fractions in the bituminous layer of Athabasca oil sands, *Energy Fuels*, 2013, 27, 4677. Copyright: American Chemical Society (2013).

A method for heavy oil liberation at the glass surface by kerosene[1] and fatty acid methyl ester (FAME[2]) has been explored and found to

---

[1] Kerosene is a low viscosity, clear liquid formed from hydrocarbons obtained from the fractional distillation of petroleum (150 - 275 °C), resulting in a mixture with a density of 0.78 - 0.81 g/cm^3 composed of carbon chains that typically contain between 10 and 16 carbon atoms per molecule.

[2] Fatty acid methyl esters are a type of fatty acid ester derived by transesterification of fats with methanol (CH_3OH).

be due to more bitumen liberation for both additives. Interestingly, the increase of salt concentration in the fluid phase exhibits a decrease in oil-water IFT and viscosity reduction. There have been several reports of sand wettability having a detrimental role based on composition: in the case of the carbonated reservoir mainly feldspar (solid soluteons of $KAlSi_3O_8$-$NaAlSi_3O_8$-$CaAl_2Si_2O_8$), sand, calcite ($CaCO_3$), magnesite ($MgCO_3$), clay, and dolomite [$CaMg(CO_3)_2$]. Various factors are present in rock wettability, which are usually based on solid composition (crystal structure, surface heterogeneity and roughness), mineral size (from 1 mm to meters), fine particles in the presence of oil component and HPHT condition (Shariatpanahi et al., 2016; Moghaddam et al., 2015). For example, clays, sulfate, and polyvalent carbonates have been discovered in Athabasca oil formations (Rao et al., 2013). Subsequently, this was the way to find particle-oil separation. Triton (Figure 2.5) as a non-ionic surfactant was used to identify key minerals such as phyllosilicates, including: calcite, kaolinite [$Al_2Si_2O_5(OH)_4$], and muscovite [$KAl_2(AlSi_3O_{10})(F,OH)_2$], as well as corundum ($Al_2O_3$) and halloysite [$Al_2Si_2O_5(OH)_4$] (Ye et al., 2008).

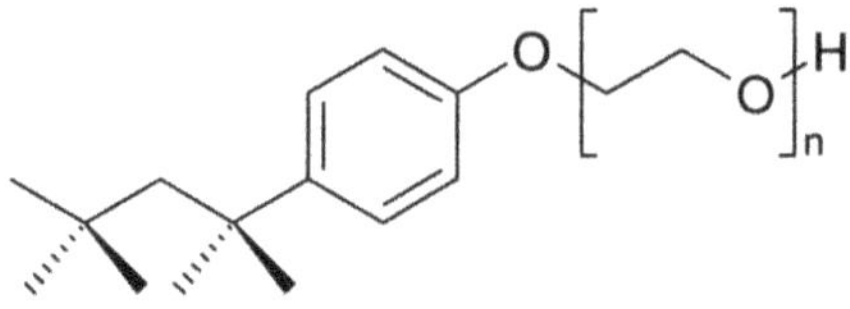

Figure 2.5: The structure of Triton X-100 nonionic surfactant.

Impact of EOR processes on rock mechanisms

The most prominent factors observed in the oil-bearing formation, causing substantial changes in sweep oil efficiencies were subsidence, compaction, stress, and strain. These behaviors occur especially for low permeability formation (shale and tight) or thick reservoirs. In this case, we lose much of the pores through the fractures by polymer diffusion and the presence of stress forces in the rocks. Moreover, high viscous flow in polymer flooding may cause negative effects such as DPR, resulting in rock expansion and an increase in rock stress (rock/formation failure.)

In principle, any changes in fluid composition will likely cause various stress behaviors in the rock reservoir. Thermal methods are faced with inter-well shear fractures as well. Furthermore, in most documented cases using gas injection (especially CO_2, and miscible gasses), geotechnical parameters are the main stress forces in rocks. For instance, after CO_2 injection, chemical dissolution between gas and fluids caused bicarbonate ion (HCO_3^-) production (Chen et al., 2015). Bicarbonate has a weak acid interaction with dissolved calcite, dolomite, and anhydrite ($CaSO_4$). This conventional dissolution tends to change the rock permeability, causing instability and CO_2 breakthrough. Understanding how these weakening interactions and stresses occur between the fluid and rock is complicated; however, the most important factors in solving this problem are the minerals types, fluid interactions, and fluid rate distribution.

Bibliography

W. Alameri, T. W. Teklu, R. M. Graves, and H. Kazemi, Wettability alteration during low-salinity waterflooding in carbonate reservoir cores, *SPE Asia Pacific Oil and Gas Conference and Exhibition, Adelaide, Australia*, 2014, SPE-171529-MS.

J. A. Ali, K. Kolo, A. K. Manshad, and A. H. Mohammadi, Recent advances in application of nanotechnology in chemical enhanced oil recovery: Effects of nanoparticles on wettability alteration, interfacial tension reduction, and flooding, *Egypt. J. Pet.*, 2018, **27**, 1371.

R. Al-Mjeni, S. Arora, P. Cherukupalli, J. van Wunnik, J. Edwards, B. J. Felber, O. Gurpinar, G. J. Hirasaki, C. A. Miller, and C. Jackson, Has the time come for EOR, *Oilfield Rev.*, 2011/2010, **22**, 16.

J. S. Buckley and T. Fan, Crude oil/brine interfacial tensions, *Petrophysics*, 2007, **48**, 175.

Y. Chen, A. S. Elhag, L. Cui, A. J. Worthen, P. P. Reddy, A. J. Noguera, A. M. Ou, K. Ma, M. Puerto, and G. J. Hirasaki, CO_2-in-water foam at elevated temperature and salinity stabilized with a nonionic surfactant with a high degree of ethoxylation, *Ind. Eng. Chem. Res.*, 2015, **54**, 4252.

S. Doorwar and K. K. Mohanty, Viscous-fingering function for unstable immiscible flows, *SPE Journal*, 2017, **22**, SPE-173290-PA.

A. E. Bayat, R. Junin, A. Samsuri, A. Piroozian, and M. Hokmabadi, Transport and retention of engineered Al_2O_3, TiO_2, and SiO_2 nanoparticles through various sedimentary rocks, *Energy Fuels*, 2014, **28**, 6255.

R. Farajzadeh, B. L. Wassing, and L. W. Lake, Insights into design of mobility control for chemical enhanced oil recovery, *Energy Rep.*, 2019, **5**, 570.

F. Gassara, N. Suri, P. Stanislav, and G. Voordouw, Microbially enhanced oil recovery by sequential injection of light hydrocarbon and nitrate in low-and high-pressure bioreactors, *Environ. Sci. Technol.*, 2015, **49**, 12594.

L. He, X. Li, G. Wu, F. Lin, and H. Sui, Distribution of saturates, aromatics, resins, and asphaltenes fractions in the bituminous layer of Athabasca oil sands, *Energy Fuels*, 2013, **27**, 4677.

F. Moeini, A. Hemmati-Sarapardeh, M.-H. Ghazanfari, M. Masihi, and S. Ayatollahi, Toward mechanistic understanding of heavy crude oil/brine interfacial tension: The roles of salinity, temperature and pressure, *Fluid Phase Equilib.*, 2014, **375**, 191.

R. N. Moghaddam, A. Bahramian, Z. Fakhroueian, A. Karimi, and S. Arya, Comparative study of using nanoparticles for enhanced oil recovery: wettability alteration of carbonate rocks, *Energy Fuels*, 2015, **29**, 2111.

A. Muggeridge, A. Cockin, K. Webb, H. Frampton, I. Collins, T. Moulds, and P. Salino, Recovery rates, enhanced oil recovery and technological limits, *Phil. Trans. R. Soc. A*, 2014, **372**, 20120320.

P. C. Myint and A. Firoozabadi, Thin liquid films in improved oil recovery from low-salinity brine, *Curr. Opin. Colloid Interface Sci.*, 2015, **20**, 105.

R. A. Nasralla, M. A. Bataweel, and H. A. Nasr-El-Din, Investigation of wettability alteration and oil-recovery improvement by low-salinity water in sandstone rock, *J. Can. Petrol. Technol.*, 2013, **52**, 144.

E. Pouryousefy, Q. Xie, and A. Saeedi, Effect of multi-component ions exchange on low salinity EOR: Coupled geochemical simulation study, *Petroleum*, 2016, **2**, 215.

M. Tagavifar, S. Herath, U. Weerasooriya, K. Sepehrnoori, and G. Pope, Measurement of microemulsion viscosity and its implications for chemical enhanced oil recovery, *SPE J.*, 2018, **23**, SPE-179672-PA.

S. Trabelsi, J.-F. O. Argillier, C. Dalmazzone, A. Hutin, B. Bazin, and D. Langevin, Effect of added surfactants in an enhanced alkaline/heavy oil system, *Energy Fuels*, 2011, **25**, 1681.

P. Zhang, M. T. Tweheyo, and T. Austad, Wettability alteration and improved oil recovery by spontaneous imbibition of seawater into chalk: Impact of the potential determining ions Ca^{2+}, Mg^{2+}, and SO_4^{2-}, *Colloids Surf. A Physicochem. Eng. Asp.*, 2007, **301**, 1.

36

Chapter 3: Surfactant Enhanced Oil Recovery

Surfactants are amphiphilic organic molecules widely used in many practical applications such as cosmetics, biological systems and petroleum processing. Increasing the flow rate of trapped oil in a reservoir lies at the center of all EOR methods (Chegenizadeh et al., 2017). Surfactant is an efficient chemical for effective oil recovery. Current leading surfactants in oil reservoirs include cationic, anionic, nonionic, bio-surfactant, and zwitterionic (Figure 3.1) (Bera et al., 2012; Daoshan et al., 2004; Sen et al., 2008; Ye et al., 2008; Zhang et al., 2015). However, there are certain drawbacks associated with the use of surfactant through the oil reservoirs, which are mainly associated with chemical interactions between the fluids and rock. In this chapter, criteria for choosing the best surfactants in the oil recovery process are considered.

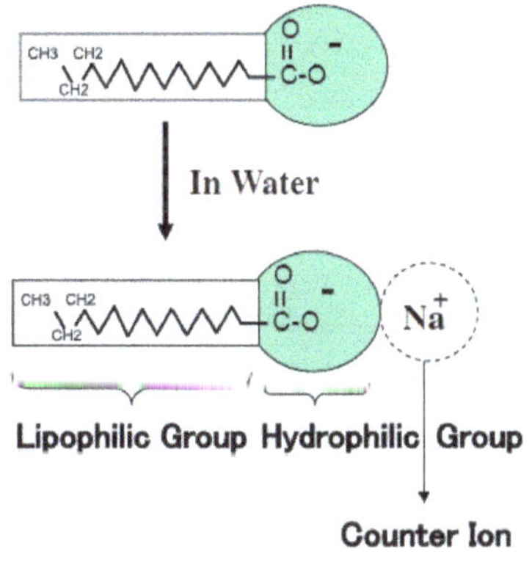

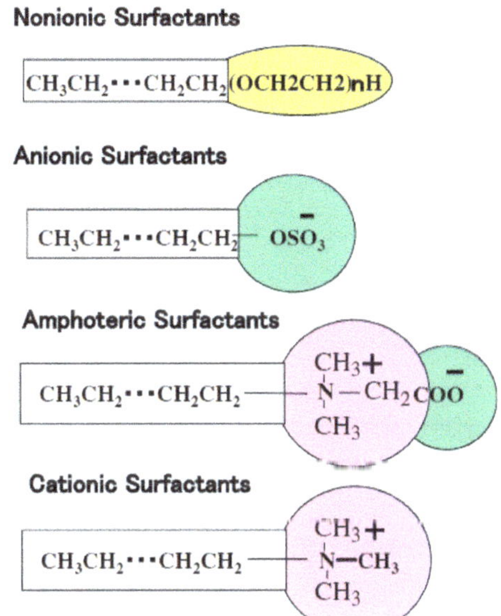

Figure 3.1: Molecular structure and classification of various surfactant molecules. Reproduced with permission from Y. Nakama, Cosmetic Science and Technology: Theoretical Principles and Applications, 2017, 15, 231. Copyright: Elsevier (2017).

Surfactant classification

What creates the biggest change at the surface/interface?

In Greek *amphi* means "from both sides" and *phile* expresses "affinity". Therefore, an amphiphilic compound depicts a double affinity of a chemical, which has polar-apolar duality (Sheng et al., 2015). A typical surfactant molecule includes a polar head and a non-polar tail. The main polar head groups are alcohol, acid, ester, phosphate, thiol, ether, sulfate, sulfonate, amine, amide, and a surfactant tail, which is an array of hydrocarbon chain including alkyl, alkylbenzene, a halogen atom, or sometimes with non-ionized oxygen atoms. The polar part shows a high affinity for polar solvents such as water (hydrophilic) and an apolar part which displays a strong affinity to less polar solvents (hydrophobic or lipophilic) (Figure 3.1). As a result of its dual affinity, the surfactant molecule is inefficient in any polar or non-polar solvent (Torres et al., 2011). Therefore, the reason that surfactant molecules tend to migrate to the surface/interface is due to the orientation of the polar head group in solvents such as water and the apolar group in oil. As a result of the new arrangement of surfactants at the interfaces, the surface tension decreases (Kiani et al., 2020). Reducing surface tension using surfactant is the basis of its use in detergent, wetting agent, soap, foaming agent, corrosion inhibitor, and antistatic agent. Surfactants are mainly categorized into three particular groups according to the charge of the head group: anionic, nonionic, and cationic (Figure 3.1).

Anionic surfactants are often considered as "the common" class of applicable amphiphiles (50% of total surfactant production). They are mainly synthesized from a number of amphiphilic anions and alkaline metal (Na^+, K^+) or quaternary ammonium. The anion in the head group can be altered for an array of negatively charged substances, creating wetting agents (di-alkyl sulfosuccinates), foaming agent (lauryl sulfates), detergent (alkylbenzene sulfonates), soaps (fatty acids), dispersants (lignosulfonates) and so on. Anionic surfactants are used frequently through the oil wells as a wetting agent to lift and suspend oil droplets in micelles (Table 3.1). In sandstone reservoirs, using anionic surfactant has been recognized as a cost-effective chemical with lower micelle adsorption. Various types of amphiphilic head groups such as sulfonate ($R-SO_3^-$), phosphate ($R-O-[PO_3]^{3-}$), sulfate ($R-O-SO_3^-$), and carboxylate ($R-CO_2^-$) can be used in EOR process. The sulfonate group has been displayed to enhance the thermal stability of the surfactant at reservoir temperature (60 - 200 °C), as well as reducing the interfacial tension (IFT), altering the rock wettability, and having a lower tendency to adsorb on the surface of sandstone reservoirs (Sheng et al., 2015; Vatanparast et al., 2011). In contrast, phosphate and sulfated groups have been shown to give low IFTs and have better performance at lower temperatures, while carboxylated groups have been developed to improve the surfactant stability at high temperatures and high salinity (Puerto et al., 2012; Negin et al., 2017).

Surfactant type	Advantages
Anionic	IFT reduction, especially applicable for sandstone reservoirs
Cationic	IFT reduction, highly stabilize in harsh condition and applicable for carbonated reservoirs
Nonionic	IFT reduction, Highly applicable in harsh reservoir condition
Zwitterionic	IFT reduction, improve wettability alteration, highly stabilize in High pressure high temperature (HPHT), suitable surfactant foam stabilizer

Table 3.1: Classes of surfactant applicable in the oilfields.

Cationic surfactants are composed of an amphiphilic cation and anions such as halogen elements (e.g., Cl, Br). The cationic head group is generally based on a nitrogen functional group, like fatty amine salts and quaternary ammonium ions, while often one or more long chains of the alkyls are available as a tail. Examples of cationic surfactants include: cetrimonium bromide (CTAB, Figure 3.2), coco alkyl trime thyl ammonium chloride {$[C_nH_{2n+1}N(CH_3)_3]Cl$ (n = 8-18)}, stearyl trimethyl ammonium chloride {$[C_{18}H_{37}N(CH_3)_3]Cl$}, dodecyltrimethylammonium bromide {DTAB, $[C_{12}H_{25}N(CH_3)_3]Br$}, and ethoxylated alkyl amine (Figure 3.3). The synthesis of anionic surfactants is much more cost-effective than cationic surfactants due to the high-pressure hydrogenation reaction which occurs during the synthesis.

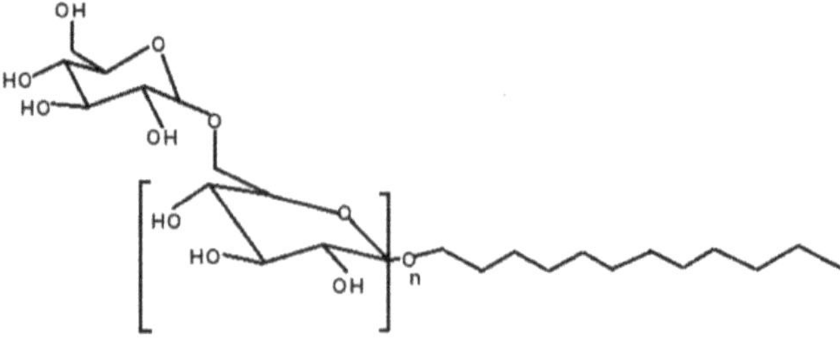

Figure 3.2: Structure of cetrimonium bromide ($[(C_{16}H_{33})N(CH_3)_3]Br$, CTAB).

Figure 3.3: Structure of ethoxylated alkylamine.

Various researchers propose the use of cationic surfactant as a good candidate to get obtain higher oil recoveries in a carbonated reservoir. The main reasons for using cationic surfactants in a carbonated reservoir are their higher stability, lower chemical interaction with rock substrate, and higher thermal stability (up to 100 °C).

A nonionic surfactant is formed during the polycondensation reaction of a polyethene glycol chain with the hydrophilic group such as alcohol, phenol, ether, ester, or amide. Once in contact with an aqueous solution, it does not ionize due to its hydrophilic group, which is a non-dissociable group. The most commonly used nonionic surfactants in the market are sugar-based (glucoside, e.g., Figure 3.4) head groups with very low toxicity.

Figure 3.4: Structure of an alkyl polyglycoside.

Designing nonionic surfactants with the ethoxy group and poly(ethylene/propylene) glycol ether makes them highly stabilized and thermally stable in high salinity, respectively.

A zwitterionic or amphoteric molecule is a single surfactant molecule, which holds simultaneously both anionic and cationic charges on its hydrophilic head group and carries a hydrophobic tail. The main properties of these structures are their low toxicity, high stability in hard water, antibacterial properties, and high compatibility with different types of surfactants (Kumar & Mandal, 2017).

What makes an exceptional oil displacement?
Typically, the residual oil saturation will be within a range of 20–30% OOIP upon contacting 100% of the given oil zone via water flooding. The oil is immobile at in this saturation range due to the surface tension between oil and water. Also, the additional pressure alone cannot overcome the high capillary pressure to move oil out from pores.

In general, surfactant leads to changes in the boundary condition of the interfaces. In general, surfactants have the lowest surface tension between surface and interfaces, amongst other materials (Figure 3.5). However, the surfactants can reduce the interfacial tension, thereby decreasing capillary pressure and allowing water to remove the trapped oil as a result of the water bypass. Furthermore, a surfactant can encourage the wettability of the reservoir as well as causing the attached oil film to lift away from the pore wall, thus decreasing residual oil saturation and enhancing oil recovery (Figure 3.6) (Zhang et al., 2009).

General principles of a surfactant molecule are decreasing IFT and changing the wetting properties of a substrate (Figure 3.5) and shifting reservoir wettability towards strongly water-wets (Figure 3.6). However, without knowing the reservoir conditions such as pressure, temperature, salinity, and rock geology, the reservoir will face unfavorable chemical reactions and blockages of pores.

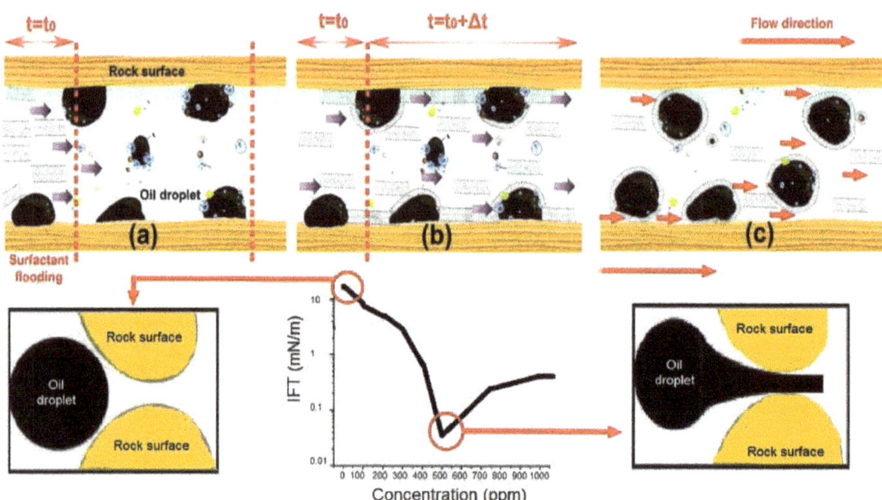

Figure 3.5: Decreasing and increasing interfacial tension (IFT). Synthesis and physiochemical characterization of zwitterionic surfactant for application in enhanced oil recovery. Adapted with permission from A. Kumar and A. Mandal, Synthesis and physiochemical characterization of zwitterionic surfactant for application in enhanced oil recovery, *J. Mol. Liq.*, 2017, 243, 61. Copyright: Elsevier (2017) and S. Kiani, S. E. Rogers, M. Sagisaka, S. Alexander, and A. R. Barron, *Energy Fuels*, 2019, 33 (4), 3162. Copyright: American Chemical Society (2019).

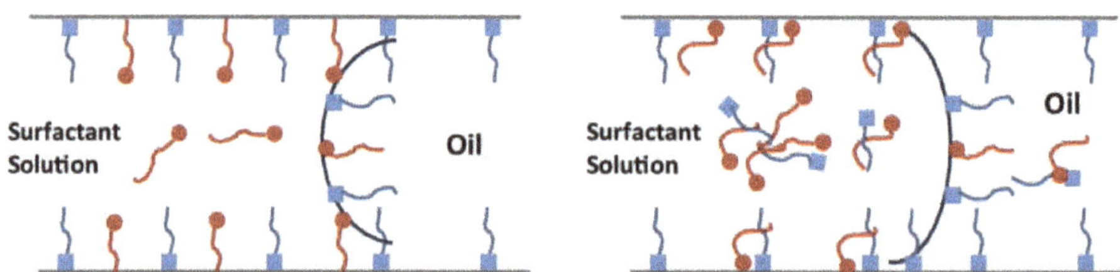

Figure 3.6: Proposed schematic of wettability alteration with (a) a non-ionic nonylphenol ethoxylate [$C_{15}H_{23}(OCH_2CH_2)_n$] surfactant and (b) a cationic cetyltrimethylammonium bromide (CTAB, $C_{19}H_{42}BrN$) surfactant. Reproduced with permission from A. Telmadarreie and J. J. Trivedi, New insight on carbonate-heavy-oil recovery: pore-scale mechanisms of post-solvent carbon dioxide foam/polymer-enhanced-foam flooding, *SPE J.*, 2016, 21, 5. Copyright: The Society of Petroleum Engineers (2016).

Asphaltene is recognized as the 'cholesterol' of crude oil within oil reservoirs (Kiani et al., 2020), and consist primarily of carbon, hydrogen, nitrogen, oxygen, and sulfur, as well as trace amounts of vanadium and nickel (Figure 3.7). The C:H ratio is approximately 1:1.2, depending on the asphaltene source. Asphaltenes are defined operationally as the n-heptane-insoluble, toluene-soluble component of a carbonaceous material such as crude oil, bitumen, or coal.

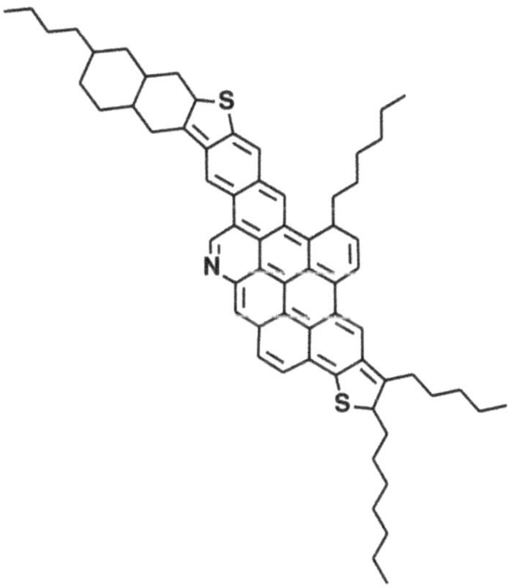

Figure 3.7: A representation structure for a component of asphaltene.

The molecular hydrocarbon component of crude oil consists of polycyclic aromatic hydrocarbon (PAH) and acidic substances found within oil-filled formations alongside aromatic hydrocarbons, resins, and saturates. Currently, many reservoirs across the world are highly viscous and the presence of asphaltene is leading to a high deposition of oil droplets on the rock surface and causing a significant decrease in the reservoir fluid pressure. In order to alleviate asphaltene aggregation, surfactant molecules can be added to increase its solubility (Figure 3.8).

The hyperbranched chains on the tail of low surface energy surfactants (LSES) cause a lowering of surface free energy and rock wettability alteration, offering significant improvement in oil recovery in asphaltene oil reservoirs. Kiani et al. reported adding hyperbranched LSES yielded a significant increase in the original-oil-in-place (OOIP) recovery (58%) relative to brine flooding (25%), even in the presence of asphaltene.

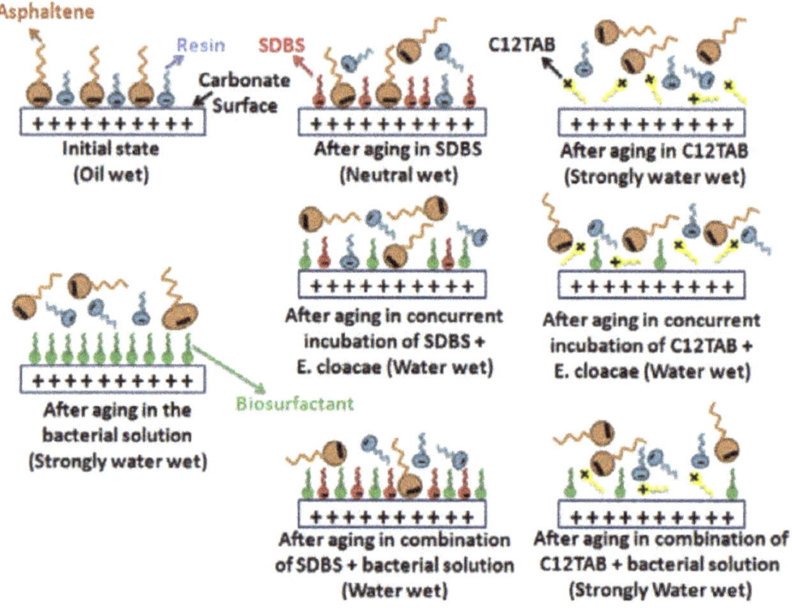

Figure 3.8: Wettability alteration in heavy oil reservoirs. Proposed interactions between rock, asphaltene, and surfactant molecules. Reproduced with permission from F. Hajibagheri, M. Lashkarbolooki, S. Ayatollahi, and A. Hashemi, The synergic effects of anionic and cationic chemical surfactants, and bacterial solution on wettability alteration of carbonate rock: an experimental investigation, *Colloid. Surf. A: Phys. Eng. Asp.*, 2017, 513, 422. Copyright: Elsevier (2017).

Main factors in a successful surfactant flooding: IFT

The term IFT is typically associated with the molecular forces that exist at the interface between two immiscible fluids. The binding energy of two immiscible fluids is largely changed by the addition of surfactant leading to a decrease in the IFT value (Figure 3.5). Besides, decreasing IFT leads to an increase in capillary forces amongst the pore spaces and causes significant oil recovery (Karnanda et al., 2013).

Temperature

Surfactant molecules are temperature-based compounds, which affects their performance in various applications. Both critical micelle concentration (CMC) and IFT are changed by temperature. Hydrophilic-lipophilic balance (HLB), cloud point, and Krafft point are three common indicators used when designing a surfactant in a particular application such as EOR. Hydrophilic-lipophilic balance (HLB) is a balance of the size and strength of hydrophilic and lipophilic moieties of a surfactant molecule (Griffin et al., 1949). In this case, HLB scales have been calculated based on Equation 3.1, where M_h and M are the molecular mass of the hydrophilic portion of the molecule and the molecular mass of the whole molecule.

$$HLB = 20 \times (M_h/M) \tag{3.1}$$

The range of the HLB scale is from 0 to 20. According to the HLB scale range, surfactants can be used in various EOR applications such as water in oil (W/O) emulsion (3-6), wetting/spreading agent (7-9), and oil in water (O/W) emulsion (8-16) (Figure 3.9) (Bera et al., 2012).

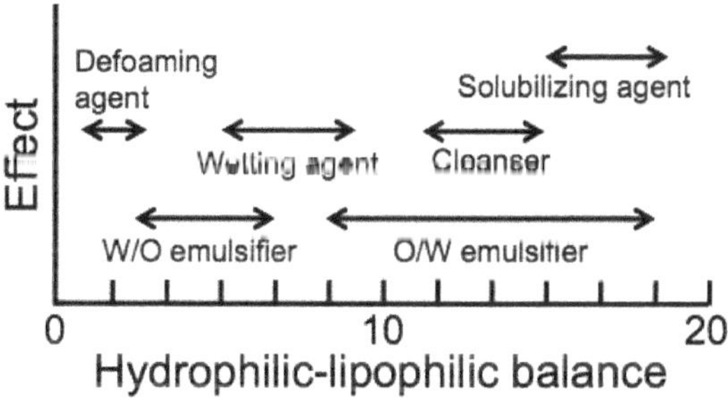

Figure 3.9: HLB scale range for choosing proper surfactant in different applications. Adapted from Y. Nakama, Cosmetic science and technology: theoretical principles and applications, *Cosm. Sci. Tech.*, 2017, 15, 231. Copyright: Elsevier (2017).

Cloud point

The temperature that surfactant separation occurs within the solution is called cloud point. Cloud point is an index to measure the solubility amount of a surfactant (Figure 3.10). Understanding the cloud point of each surfactant is critical in preventing molecule aggregation.

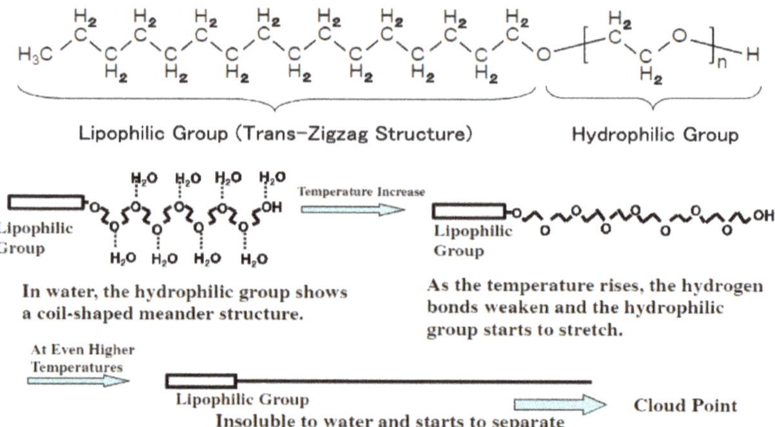

Figure 3.10: Effect of increasing temperature in cloud point of polyoxyethylene surfactants. Reproduced with permission from Y. Nakama, Cosmetic science and technology: theoretical principles and applications, *Cosm. Sci. Tech.*, 2017, 15, 231. Copyright: Elsevier (2017).

Krafft point

The Krafft point is defined as the point at which surfactant reaches its highest solubility in water. The surfactant solution at a lower Krafft point would form a bilayer structure of hydrated solid and in higher Krafft point, would form micelles. Molecular structures of surfactants have the main effect in the precise determination of the Krafft point (Figure 3.11) (Glover et al., 1979).

Brine effect on surfactant

The solubilization potential of amphiphilic substances follows their structure, inorganic salts, and temperature. Typically, long hydrophobic groups obtain the maximum surfactant solubilization; however, surfactants are highly sensitive to the addition of salt and it suppresses the electrostatic repulsive forces between the hydrophilic head groups (Co et al., 2015, Choi et al., 2014).

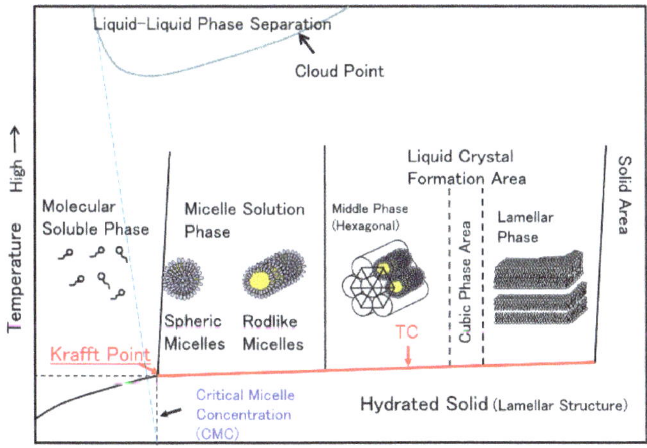

Figure 3.11: The cloud point, Krafft point, and CMC value in the concentration-temperature curve. Reproduced with permission from Y. Nakama, cosmetic science and technology: theoretical principles and applications, *Cosm. Sci. Tech.*, 2017, 15, 231-244. Copyright: Elsevier (2017).

Ideal surfactant

Many efforts have been made to create the ideal surfactant system with high synergistic properties to effectively push oil out. However, using a particular surfactant in field-scale is not an effective solution to heal oil transport through the pore space and should be used via alternative chemicals such as polymers and alkaline. A successful surfactant should ideally be highly salt, tolerant, possess high thermal stability, polymer compatibility, and be cost-effective for the EOR process (Co et al., 2015; Puerto et al., 2012; Fletcher et al., 2015).

Branched surfactants

Oil displacement by branched surfactant has proven to be a potential remedy to the difficulties in oil sweeping and fluid flow viscous fingering. For example, a branched C_{16-17} propoxy sulfate is stable in high-pressure high temperature (HPHT), resulting in more than 90% oil recovery. Despite this result, two major problems still remain in surfactant flooding: expense and environmental impact (Chegenizadeh et al., 2017).

Foam EOR scenario

Foam flooding

Fluid displacement is "hard" geological work, where more than 60% of oil remains unproduced due to the unfortunate decline in natural forces and pressure. Enhancing oil displacement with chemicals, especially foam, where the appropriate surfactant is selected, is shown to be a potential solution in addressing these problems (Li et al., 2019; Sun et al., 2014).

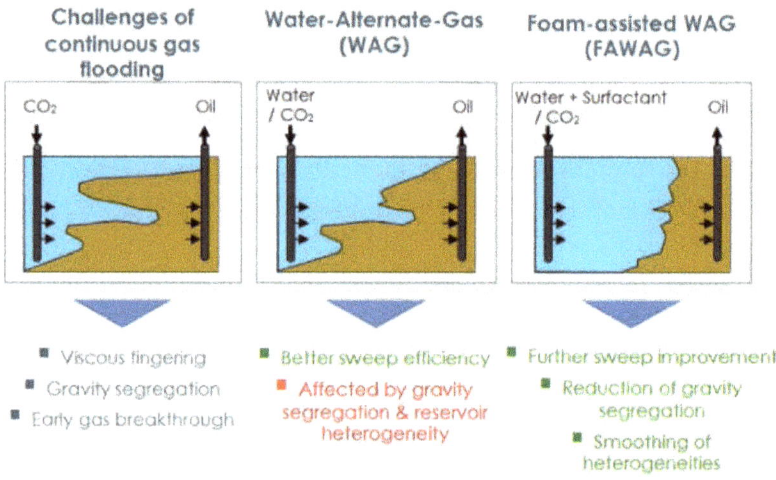

Figure 3.12: Comparison of the gas, water alternate gas (WAG), and foam assisted WAG flooding across the oilfield. Reproduced with permission from M. Sagir, M. Mushtaq, M. S. Tahir, M. B. Tahir, N. Abbas, and M. Pervaiz, CO_2 Capture, storage, and enhanced oil recovery applications, *Encyclopedia of Renewable and Sustainable Materials*, 2018, 3, 52. Copyright: Elsevier (2018).

Foam is the dispersion of a high volume of gas (99 wt.%, N_2, CO_2, and air) into a small volume of liquid (1 wt.%). Gas purging and entrapment, within surfactant solution, leads to the adsorption of surfactant molecules into the gas-liquid interface, and as a result, this generates a continuous foam bubble with a thin-stable liquid/gas interface called lamellae.

The effectiveness of foam in porous media is far greater than water and gas. This leads to reduced gas relative permeability by bubble trapping and causes an increase in the gas viscosity, which provides stable and robust foam (Figure 3.12) (Gonzenbach et al., 2006). Furthermore, foam can potentially help to improve oil sweep efficiency and reduce gas mobility by forcing more gas from a small to a highly penetrable area (Telmadarreie & Trivedi, 2016). Figure 3.12 provides a comparison between continuous gas, water-alternate-gas (WAG), and foam-assisted water alternate gas injection in the oil reservoirs.

Foam stability

Stability is one of the most important characteristics of foam. Foam coalescence occurs via three different mechanisms. The first being liquid drainage as a result of gravity force, the second is coarsening of foam caused by gas diffusion among bubbles which generates differing capillary pressures (lamellae film getting thinner), and lastly, the bubble coalescence, which occurs through gas diffusion from smaller bubbles with high pressure to larger bubbles with low pressure amid neighboring bubbles. This leads to reduced foam numbers by gravity and capillary forces, breakage in lamellae to plateau border and merging of the two smaller bubbles, forming a larger bubble. The rate of liquid drainage and critical film thickness is dependent on the surfactant charge-concentration, surface-liquid viscosity, and surface elasticity.

Impact of oil in foam stability

Oil has also been shown to have a detrimental effect on foam stability. Various studies have shown the difficulty in generating foam and obtaining stability in the presence of different types of oil. However, this is a highly contradictory subject in terms of the results obtained by various researchers. It has been demonstrated that the use of light oil causes the collapse of foam formation and generation. Other researchers believe stable foam in medium and heavy oil is a result of using the proper surfactant. It has been demonstrated previously that the use of light oil alkene chain hydrocarbons causes a drastic decrease in the half-time stability of a single foam bubble, this demonstrates the direct relationship between oil type and foam stability. Clearly, the same approach for long-term pseudo-stabilized foam film can be observed with inherent physicochemical interactions of surfactant in foam-oil interfaces. The underlying process for these interactions amongst oil-bubble can be described in three steps: the first being access of oil to the interface, then spreading on the interface, and finally creating an unstable bridge across the lamellae-thin film.

Surfactant impact on foam generation

Surfactant has a key role in stabilizing foam. The presence of surfactant creates a new layer on the fluid-fluid and solid-fluid interfaces via new chemical interactions such as van der Waals, π-π, electrostatic, steric, acid-base, etc. Therefore, enhancing surfactant diffusion on the interfaces is shown to facilitate oil separation (decreasing IFT) thus reducing the capillary pressure forces (Pc). However, as a result of large surface areas, their efficiency is inadequate due to foam instability and coalescence of bubbles in oil (seminal problems). Therefore, researchers' efforts have been focused on generating thermodynamically stable foam using surfactants, nanoparticles, and polymers in an effort to improve oil displacement with stabilized foam (Ali et al., 2018; Tu et al., 2013).

Based on these results, most foam stability is controlled by surface tension, IFT, viscoelastic behavior, composition and various types of chemicals, namely surfactants (Chegenizadeh et al., 2017). However, regarding abovementioned analyses, further studies are needed to obtain a full understanding of monomer arrangements and their structures within a solution. This provides clear evidence that the surfactant formation creates a strong thin film around bubbles in the foam.

Clearly, small-angle neutron analysis (SANS) can distinguish between aggregation and formation of surfactant in order to see the potential formation of a thin film. Figure 3.13 shows the comparison of CO_2 EOR and CO_2-foam flooding. Foam leads to decrease CO_2 gas fingering across the oilfield (Telmadarreie & Trivedi, 2016).

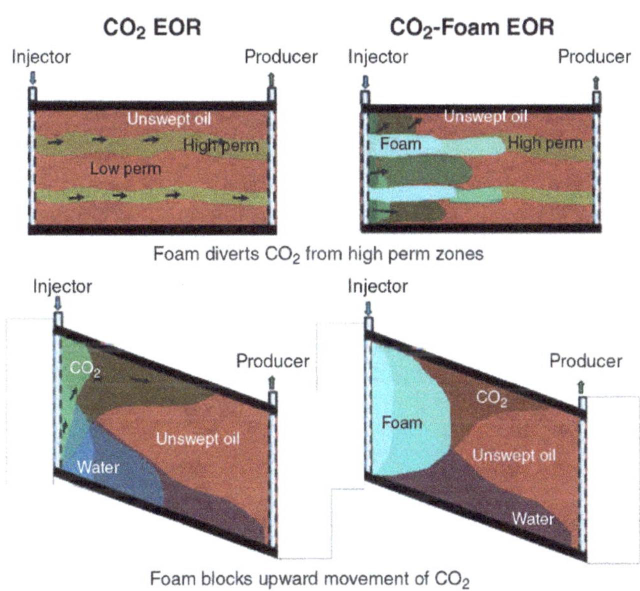

Figure 3.13: Comparison of the gas and foam flooding. Foam leads to decrease CO_2 gas fingering across the oilfield. Reproduced with permission from M. Sagir, M. Mushtaq, M. S. Tahir, M. B. Tahir, N. Abbas, and M. Pervaiz, CO_2 capture, storage, and enhanced oil recovery applications, *Encyclopedia of Renewable and Sustainable Materials*, 2018, 3, 52. Copyright: Elsevier (2018).

Surface and interfacial tension (IFT) effect on foam stability

It is generally believed that optimum surface and interfacial tension values have a positive effect on foam stability. It seems there is a direct link between the charges on the head of surfactants and foam generation-stabilization. This provides an effective thin film around the bubble. In this case, selecting a good surfactant results in homogenous stretching forces in a thin film. This increases foam stabilization by reducing the local surface tensions (Gibbs energy) amongst bubbles. It is generally believed that optimum surface and interfacial tension values have a positive effect on foam stability. Over the past decade, different viscoelastic surfactants (VES) were synthesized and developed to increase the viscosity of solutions, as well as control the rheological properties in different industrial applications e.g., drilling fluid, hydraulic fracturing, and EOR methods (Chegenizadeh et al., 2017).

Surfactant type and size

Additionally, surfactant size needs further investigation in order to observe the stability and aggregations to generate a homogenous foam displacement system. Using this information, we can see how surfactants size and stabilities can have a significant effect on foam displacement, a piston-like effect, even. The composition and charges of surfactant in foam are intended to make homogenous foam fronts. This greatly affects the adsorption of monomer layers between bubble, surfactant, and oil. Thus, creating higher external fluid displacement. In addition to this, it is observed that the type of surfactant used i.e., the charge and composition, can affect foam systems, in that it creates a weak or strong interaction.

Marangoni flow and bubble coalescence

In foam systems, Marangoni flow depends on the surface elasticity within foam solutions.[1] Surface elasticity has a dominant effect, leading to a decrease in the stability of thin liquid films. By selecting the right viscoelastic surfactant, bubble coalescence (film rupture) by Ostwald ripening can be rendered. However, this cannot distribute homogeneously over entire foam bubble surfaces. This is due to insufficient monomer around the foam bubble, resulting in local resistance to extension. This generates a flow against local surface forces and toward the depleted areas (this takes place after stretching) called Marangoni flow (Mellema & Benjamins, 2004). In the foam systems,

---

[1] The Marangoni effect (also called the Gibbs–Marangoni effect) is the mass transfer along an interface between two fluids due to a gradient of the surface tension.

Marangoni flow relates to surface elasticity within foam solutions. In the foam systems, surface elasticity has a dominating effect leading to a decrease in the stability of thin liquid films (Figure 3.14). By selecting a good viscoelastic surfactant, bubble coalescence (film rupture) by Ostwald ripening can be rendered at low elasticity. Furthermore, there is a positive relationship between surfactant solution, its concentration, and foam generation with both viscosity and shear stress factors in bulk solution (Mellema & Benjamins, 2004).

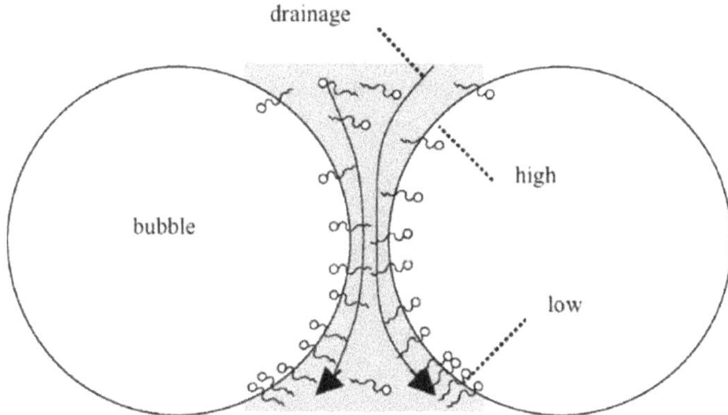

Figure 3.14: Schematic of Marangoni-effect between two bubbles in a foam. Reproduced with permission from M. Mellema, J. Benjamins, Importance of the Marangoni effect in the foaming of hot oil with phospholipids, *Colloid Surf. A: Physicochem. Eng. Asp.*, 2004, 237, 113. Copyright: Elsevier (2004).

Bibliography

A. Bera, K. Ojha, T. Kumar, and A. Mandal, Water solubilization capacity, interfacial compositions and thermodynamic parameters of anionic and cationic microemulsions, *Colloid Surf. A Physicochem. Eng. Asp.*, 2012, **104**, 70.

N. Chegenizadeh, A. Saeedi, and X. Quan, Most common surfactants employed in chemical enhanced oil recovery, *Petroleum*, 2017, **3(2)**, 197.

L. Co, Z. Zhang, Q. Ma, G. Watts, L. Zhao, and P.J. Shuler, Evaluation of functionalized polymeric surfactants for EOR applications in the Illinois basin, *J. Petroleum Sci. Eng.*, 2015, **134**, 167.

K.-O. Choi, N.P. Aditya, and S. Ko, Effect of aqueous pH and electrolyte concentration on structure, stability and flow behavior of non-ionic surfactant based solid lipid nanoparticles, *Food Chem.*, 2014, **147**, 239.

L. Daoshan, L. Shouliang, L. Yi, and W. Demin, The effect of biosurfactant on the interfacial tension and adsorption loss of surfactant in ASP flooding, *Colloid Surf. A Physicochem. Eng. Asp.*, 2004, **244**, 53.

C. Glover, M. Puerto, J. Maerker, and E. Sandvik, Surfactant phase behavior and retention in porous media, *SPE J.*, 1979, **19**, 183.

F. Hajibagheri, M. Lashkarbolooki, S. Ayatollahi, and A. Hashemi, The synergic effects of anionic and cationic chemical surfactants, and bacterial solution on wettability alteration of carbonate rock: An experimental investigation, *Colloid. Surf. A: Phys. Eng. Asp.*, 2017, **513**, 422.

G. Hirasaki, C.A. Miller, and M. Puerto, Recent advances in surfactant EOR, *SPE J.*, 2011, **16**, 889.

W. Karnanda, M.S. Benzagouta, A. Al-Quraishi, M.M. Amro, Effect of temperature, pressure, salinity, and surfactant concentration on IFT for surfactant flooding optimization, *Arab. J. Geosci.*, 2013, **6**, 3535.

S. Kiani, D. R.Jones, S. Alexander, and A. R. Barron, New insights into the interactions between asphaltene and a low surface energy anionic surfactant under low and high brine salinity, *J. Colloid Interf. Sci.*, 2020, **571**, 307.

S. Kiani, S. E. Rogers, M. Sagisaka, S. Alexander, and A. R. Barron, A new class of low surface energy anionic surfactant for enhanced oil recovery, *Energy Fuels*, 2019, **33**, 3162.

A. Kumar and A. Mandal, Studies on interfacial behavior and wettability change phenomena by ionic and nonionic surfactants in presence of alkalis and salt for enhanced oil recovery, *J. Mol. Liq.*, 2017, **243**, 61.

M. Mellema, J. Benjamins, Importance of the Marangoni effect in the foaming of hot oil with phospholipids, *Colloid Surf. A: Physicochem. Eng. Asp.*, 2004, **237**, 113-118.

Y. Nakama, Cosmetic Science and Technology: Theoretical Principles and Aplications, *Cosm. Sci. Tech.*, 2017, **15**, 231.

C. Negin, S. Ali, and Q. Xie, Most common surfactants employed in chemical enhanced oil recovery. *J. Petro.*, 2017, **3**, 197-211.

M. Sagir, M. Mushtaq, M. S. Tahir, M. B. Tahir, N. Abbas, and M. Pervaiz, CO_2 Capture, Storage, and Enhanced Oil Recovery Applications, *Encyclopedia of Renewable and Sustainable Materials*, 2018, **3**, 52.

R. Sen, Biotechnology in petroleum recovery: the microbial EOR, *Prog. Energy Combust. Sci.*, 2008, **34**, 714.

J. J. Sheng, Status of surfactant EOR technology, *Petroleum*, 2015, **1**, 97.

L. Torres, A. Moctezuma, J.R. Avendaño, A. Muñoz, and J. Gracida, Comparison of bio-and synthetic surfactants for EOR, *J. Pet. Sci. Eng.,* 2011, **76**, 6.

A. Telmadarreie and J. J. Trivedi, New insight on carbonate-heavy-oil recovery: pore-scale mechanisms of post-solvent carbon dioxide foam/polymer-enhanced-foam flooding, *SPE J.,* 2016, **21**, 651.

H. Vatanparast, A.H. Alizadeh, A. Bahramian, and H. Bazdar, Wettability alteration of low-permeable carbonate reservoir rocks in presence of mixed ionic surfactants, *Pet. Sci. Tech.,* 2011, **29**, 1873.

Z. Ye, F. Zhang, L. Han, P. Luo, J. Yang, and H. Chen, The effect of temperature on the interfacial tension between crude oil and gemini surfactant solution, *Colloid Surf. A Physicochem. Eng. Asp.,* 2008, **322**, 138.

J. Zhang, Q.P. Nguyen, A. Flaaten, G.A. Pope, Mechanisms of enhanced natural imbibition with novel chemicals, *SPE J.,* , 2009, **12**, 912.

Q.-Q. Zhang, B.-X. Cai, W.-J. Xu, H.-Z. Gang, J.-F. Liu, and S.-Z. Yang, Novel zwitterionic surfactant derived from castor oil and its performance evaluation for oil recovery, *Colloid Surf. A Physicochem. Eng. Asp,* 2015, **483**, 87.

56

Chapter 4: Polymer Enhanced Oil Recovery

The production of heavy oil using a polymer is a highly plausible process in fossil reserves. For an economical EOR method, polymers that are sufficiently stable, cost-effective, and able to increase displacing fluid rheology is required to enhance sweep efficiency ($M < 1$) (Firozjaii & Saghafi, 2019). In this chapter, we show that injection of various polymers at high-volume creates an ideal treatment to overcome fluid channeling via the increase of viscosity of displacing fluid. We will also briefly touch upon polymer chemistry, an optimal polymer composition, examples of successful polymer flooding, and oil demulsification, using polymers.

Polymer flooding

What is polymer?

A polymer is a large molecule structure composed of many smaller repeating chemical subunits, known as monomers (Firozjaii & Saghafi, 2019). In Greek *poly* means "many" and *mer* is "unit". Polymers pose beneficial features such as molecular weight (ranging from few thousand to several million Daltons). According to reservoir features such as formation rock, formation water, and ion concentration, applicable polymers in an EOR process are within a range of 3-35 million Dalton. Polymers are comprised of repeat units named according to the monomer applied. For example, polyacrylamide contains repeating units derived from acrylamide (Figure 4.1). Polymers are categorized into two groups:

+ natural polymers,
- synthetic polymers.

The first being synthetic, such as hydrolyzed polyacrylamide (HPAM, Figure 4.1), the most frequently used polymer in reservoirs, and the second is a biopolymer, which consists of biomaterials such as non-ionic xanthan gum (Figure 4.2). Adding water-soluble polymers to the base-fluid increases its' viscosity, which aids displacement of homogenized fluid through the formation; known as polymer flooding. An anionic polymer can expand in a polar medium (water) due to its electrostatic repulsions. There is a positive correlation between how large the hydrodynamic volume is and the resulting viscosity measurements. The use of anionic polymers minimizes adsorption in primarily sandstone reservoirs and has been common practice since the late 1960s. Over 80% of real polymer flooding projects such as Daqing, Mangala, and Pelican Lake used HPAM. The main advantage of HPAM flooding is its cost-effectiveness, good adsorption on the rock substrate, and high fluid viscosity (shear thickening). Towards the end of the 1980s, the polymer project, which used HPAM was uneconomical ($\sim$ \$16/bbl), therefore unfavorable for use in the reserves. That being said, there are limitations to the applicability of HPAM such as in the event of acrylamide becoming hydrolyzed and converting to acrylate groups when at high temperatures (above 60 °C), and as a consequence HPAM becomes precipitated.

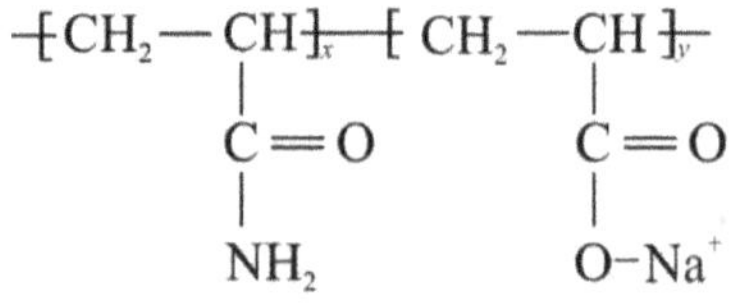

Figure 4.1: The structure of hydrolyzed polyacrylamide (HPAM).

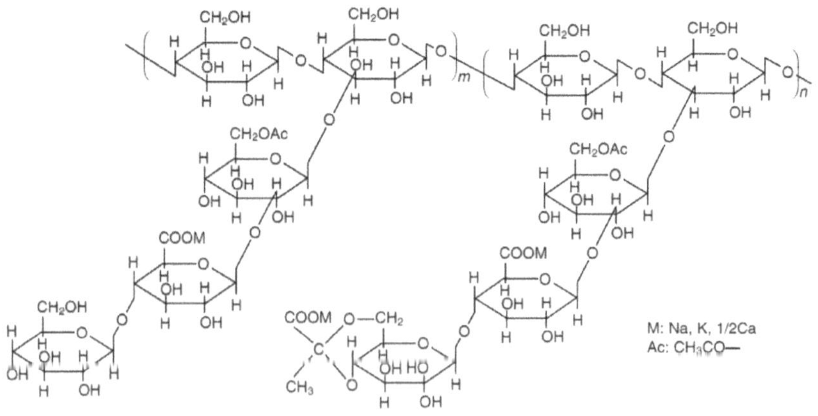

Figure 4.2: The chemical structure of xanthan gum.

Some examples of applicable polymers

Several polymers such as biodegradable and synthetic polymers have been used based on a suitably functionalized series of a biodegradable monomer such as acrylamide (Figure 4.3a), acrylic acid (Figure 4.3b), konjac glucomannan (Figure 4.4), and allyl polyoxyethylene (Figure 4.5).

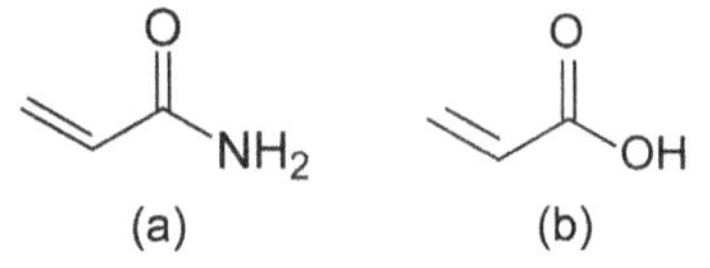

Figure 4.3: Structures of (a) acrylamide (AM) and (b) acrylic acid.

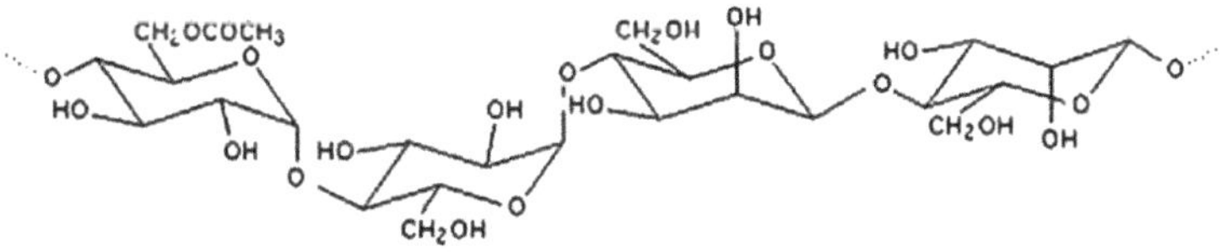

Figure 4.4: The chemical structure of the water-soluble polysaccharide konjac glucomannan.

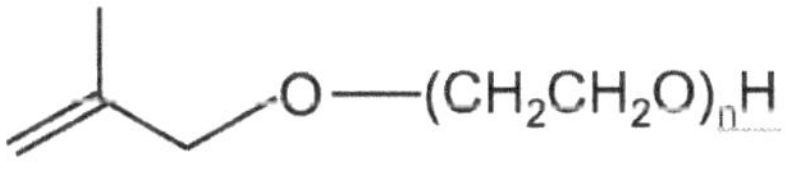

Figure 4.5: Structure of allyl polyoxyethylene.

When the above-mentioned polymers are mixed with the brine solution at rock formation, increase the viscosity of displacing fluid, reduce the relative permeability, and finally increase the oil sweep efficiency. The most common anionic polyacrylamides used in the EOR industry are copolymers of acrylamide and acrylic acid. Acrylamide is a white, crystalline, water-soluble compound derived from acrylonitrile. It contains an electron-deficient double bond and an amide group and undergoes chemical reactions typical of these two functionalities. Also, acrylic acid is a moderately strong carboxylic acid. They exhibited good viscoelasticity and shear resistance, high stability in harsh condition (6700 mg/L salt solution at 65 °C), and proper biodegradability (degradation rate > 60%) (Firozjaii & Saghafi, 2019). Although these are favorable properties, they still give rise to the following questions: how do we identify all of the biodegradable microorganisms in the polymer/brine/oil system? What combination of microorganisms lead to the polymer degradation? What type of biocide prevents the polymer degradation?

Polyoxyethylenes (Figure 4.6), which are an ethylene glycol group made from various ethylene glycol units. Polyoxyethylenes are a unique emulsion stabilizer, which can be used in saltwater over a long period.

$$HO\left[\!\!\!\diagdown\!\!\!\diagup_O\right]_n\!\!\!H$$

Figure 4.6: Structure of polyoxyethylene.

ATBS (or AMPS™) stands for acrylamido tertiary butyl sulfonic acid (or, strictly, acrylamide-2-methylpropane sulfonic acid). ATBS is made by the Ritter reaction of acrylonitrile and isobutylene in the presence of sulfuric acid and water (Figure 4.7). This monomer was studied to overcome the stability issues in amide groups at high temperatures. The dimethyl and sulfomethyl ($-CH_2SO_2OH$) groups sterically hinder the amide function and provide thermal and hydrolytic stability to ATBS-containing polymers. Excessive hydrolysis of amide groups to carboxylate is a major cause of instability of polyacrylamides; it can cause polymer precipitation and, in turn, viscosity loss. At higher temperatures and/or when the divalent content increases, it is possible to incorporate ATBS to prevent precipitation and extend the temperature range of application. Numerous studies have been performed to evaluate the benefits of incorporating this sulfonated monomer; the main result is that ATBS increases salinity and temperature tolerance up to 95 °C. Additional studies have shown that the presence of ATBS increases shear stability and decreases retention in reservoirs.

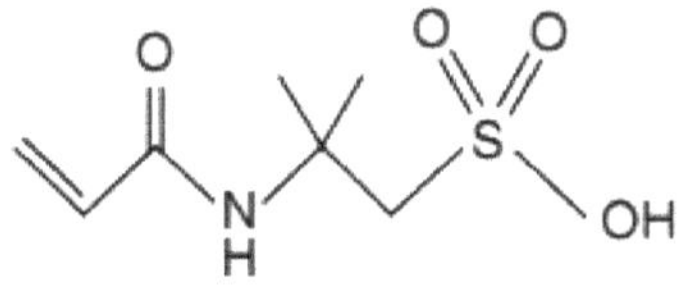

Figure 4.7: Structure of acrylamido tertiary butyl sulfonic acid (ATBS).

N-vinylpyrrolidone (NVP, Figure 4.8) is an organic compound produced from the reaction of acetylene with 2-pyrrolidone. This monomer can protect the neighboring acrylamide groups from hydrolysis and improves temperature stability; however, the reactivity of this monomer is relatively low, resulting in composition drifts and low molecular weights. For higher temperatures, it is possible to add NVP to protect acrylamide from hydrolysis, as discussed by several authors. The temperature limit for such polymers can be extended up to 140°C in certain conditions. However, these polymers have lower molecular weights than conventional copolymers of acrylamide and acrylic acid, for instance, due to reactivity issues during polymerization. This often requires a higher polymer dosage to reach a given viscosity, thus leading to higher costs.

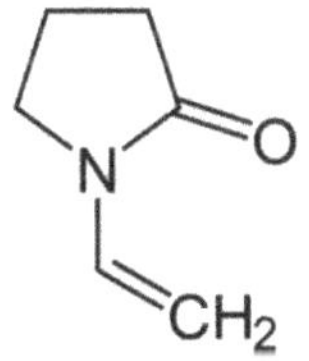

Figure 4.8: Structure of N-vinylpyrrolidone (NVP).

Thermo-viscosifying polymers (TVPs) are a class of novel synthetic polymers, which stabilize at high temperatures and high salinity. Compared to other water-soluble polymers, TVPs create better fluid viscosity and elastic modulus at high temperatures. Therefore, they are useful chemicals to be used in the reserves. For example, polyacrylamide, as a TVP, can be stabilized at higher temperatures within electrolyte solutions via the sulfation method. Another class of water-soluble synthetic polymers is temperature-responsive polymers that alter with temperature. These alterations depend on their chains and side-groups through lower critical solution temperature (LCST) moieties and upon reaching optimal temperature will respond accordingly. Polymers that increase viscosity with temperature are advantageous in EOR applications. Firstly, injection of a polymer solution with low viscosity is optimal for short-term flooding, and secondly their ability to retain mobility control at high temperatures in saline medium.

What does a polymer solution do in pore spaces?

The addition of polymer to fluid solution leads to enhance fluid viscosity (Abdulbaki et al., 2014). By increasing the viscosity of displacing fluid, rock permeability reduces, and fluid mobility can be better controlled. As we are aware, polymers applied in EOR process have molecular weights ranging from 3 - 35 million Daltons. The viscosity measurement of polymers is usually calculated according to molecular weight. The transfer agent controls the final range of molecular weight during the polymerization stage. Therefore, designing polymers with optimum molecular weight is a critical factor in achieving successful oil recovery, for if they are too large in size, they block the pores. For instance, the experiments have generally shown that pore plugging will occur when the size of the polymer exceeds one-fifth of the pore throat size in the oil formation. Very few numbers of solubilizing polymers can be used in the reservoirs, which significantly affects both sweep efficiency and the mobility ratio (Firozjaii & Saghafi, 2019).

An optimal composition

A suitable polymer in a harsh environment has two remarkable properties: (a) good viscosity with minimal adsorption on the rock substrate (b) good thermal stability in contact with saltwater at a higher temperature (up to 150 °C). During the past 50 years, various polymer structures have been developed as an EOR agent for oil recovery (Figure 4.9). However, the majority of polymers cannot be used in harsh environments due to their structural properties. First of all, the backbone (carbon chain) structure has a critical role in the thermal stability of polymers. Lacking -O- bonding in the backbone increases the success rate of polymer flooding (Figure 4.9).

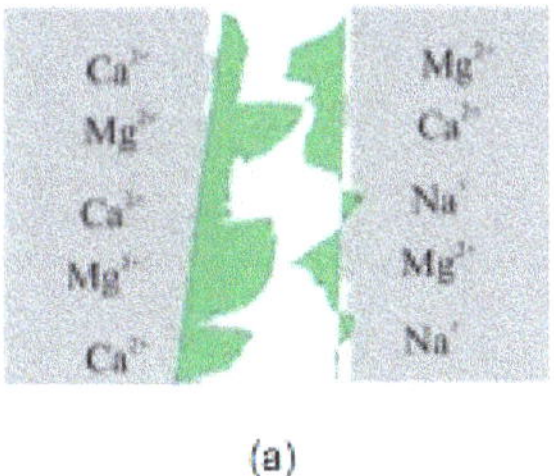

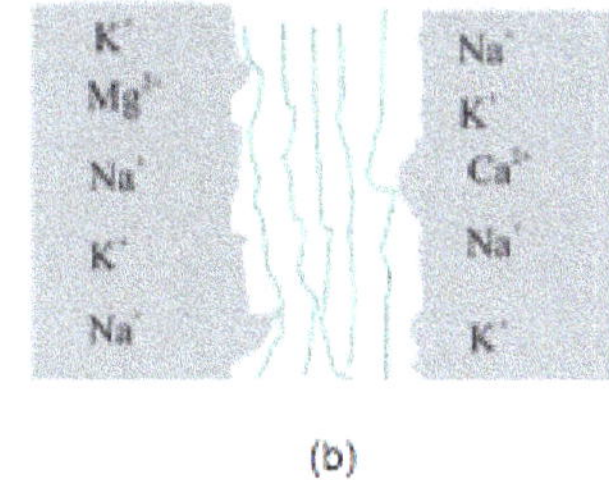

Figure 4.9: Polymer flooding across the wellbore showing (a) the polymer molecules adsorption by divalent ions and permeability reduction at the high-temperature-high-shear (HTHS) condition, and (b) polymer molecules movement in porous medium at the low salinity low-temperature-low-shear (LTLS) condition. Reproduced with permission from A. M. Firozjaii and H. R. Saghafi, Review on chemical enhanced oil recovery using polymer flooding: fundamentals, experimental and numerical simulation, *Petroleum*, 2020, 6, 115.

An increase in a degree of hydrolysis in relation to temperature and time is often what is behind thermal degradation (molecule precipitation) of water-soluble polymers in harsh environments. The primary goal of polymer flooding is retaining viscosity of solution through optimized chemistry. The correct selection of polymers with proper structure reduces the possibility of backbone degradation in higher temperatures. Typically, the risk of thermal degradation (above 80 °C) is enormous when an -O- bond is in the backbone. For instance, sodium alginate (Figure 4.10), xanthan gum (Figure 4.2), carboxymethyl cellulose (Figure 4.11), and polyoxyethylene (Figure 4.6) have shown lower thermal stability at high temperatures due to the presence of -O- bonding in their backbone. Secondly, the presence of negatively ionic hydrophilic groups leads to reduced adsorption of polymer on the rock surface. HPAM (Figure 4.1), polyvinyl, polyacrylamide (Figure 4.3) and sodium polyacrylate (Figure 4.12) are among the most common examples of polymers with low thermal degradation and high stability.

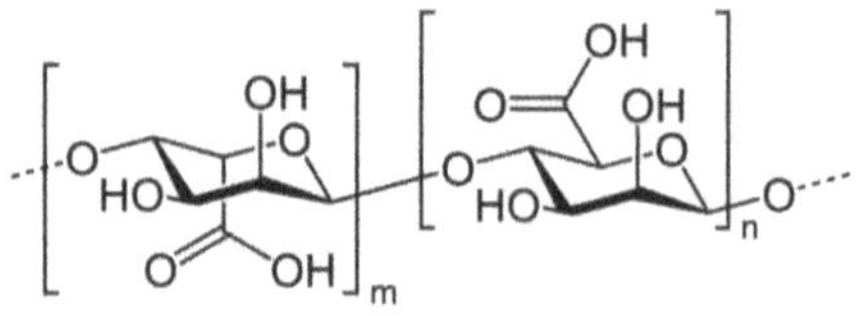

Figure 4.10 Structure of alginic acid.

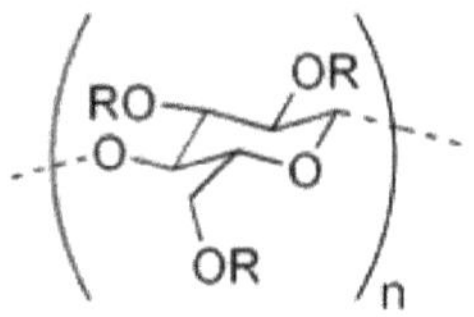

Figure 4.11 Structure of carboxymethyl cellulose: R = H or CH_2CO_2H.

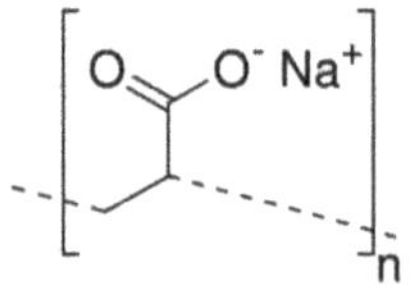

Figure 4.12 Structure of sodium polyacrylate, also known as waterlock.

Generally, the following conditions are known to result in polymer degradation: formation rock type, water salinity, reservoir temperature, polymer composition, surfactant compatibility, and the formation rock wettability/permeability/pore structure. The addition of appropriate functionalized groups to the polymer backbone is an absolute must in order to decrease polymer retention, enhance fluid viscosity, and maximize the reservoir pressure gradient. For instance, minimal polymer adsorption in a wide range of reservoir conditions was observed after adding $-CO_2^-$ as a hydrophilic group, however, the $-CO_2^-$ group is mainly salt-dependent (divalent cations such as Mg^{2+} and Ca^{2+}). To increase polymer stability, using -OH or $-CONH_2$ bonding in the hydrophilic group decreases repulsion forces amongst chain links even after the addition of divalent cations. Therefore, future synthesis should focus on optimizing polymers using -COO, -OH, and -$CONH_2$ to improve the water viscosity and in turn, improve oil mobility in real reservoirs. Table 4.1 shows successful polymer flooding using various amounts of HPAM. The oil production rate grows with increasing HPAM concentration. However, the fact remains that the oil recovery factor depends on many factors such as rock geology, permeability, salinity, and fluid viscosity.

Polymer conc. (ppm)	Polymer viscosity (cp)	Oil viscosity (cp)	Lithography, porosity (%)	Permeability (mD)	Oil recovery (OOIP, %)
500	25	209	Sandpack, 34.8%	4500	32.5%
250	9	1140	Sandpack, 35.1%	5300	36.3%
1500-2000	50-100	2000	Sandpack, 32%	1760	15%
5644	76	1450	Sandpack, 35%	7000	19%
2427	15.5	1450	Sandpack, 35%	7000	19.8%
2895	22	1450	Sandpack, 35%	7000	20.9%
1500	25	1600	Sandpack, 40%	5700	22%
10000	n/a	1000	Sandpack, n/a	13000	27.2%
10124	360	5500	Sandpack, 35%	7000	21%
1500	25	950	Sandstone, 30%	1856	25%
1000	50	764	Synthetic core, n/a	12000	44%
4000	29	203	Sandpack, 32.8%	2350	12.9%
1650	62	7000	Sandstone, 22%	2450	29.8%
n/a	55	1904	Sandpack, n/a	4035	34.2%
1500	85	18700	Sandpack, 39.6%	2825	39 %

Table 4.1: Experimental HPAM flooding in heavy oil reservoirs. Data reproduced from H. Saboorian-Jooybari, M. Dejam, and Z. Chen, Heavy oil polymer flooding from laboratory core floods to pilot tests and field applications: Half-century studies, *J. Pet. Sci. Eng.*, 2016, 142, 85. Copyright: Elsevier (2016).

Polymer versus water flooding

Polymer flooding has three main benefits when compared to water flooding;
- creates a homogenous fluid displacement,
- lowers mobility ratio of water to oil,
- improves the sweep efficiency.

On the other hand, when water alone is injected into the reservoir, it surpasses the oil and leads to fluid viscous fingering. This phenomenon occurs when a fluid such as water with low viscosity diffuses faster than oil, consequently leading to large amounts of oil remaining throughout the pores and pore-throats (Table 4.2).

Oilfield	Rock porosity (%)	Rock permeability (mD)	Formation temp. (°C)	Oil gravity (API)	Oil recovery (%)
Court Bakken	29	2100	30.8	17	30
Lloydminster	32	2000	22.2	15	1-2
Battrum	25	1300	37.2	16	6-9
Buffalo	24	767	25	13	4.1
Karamay	20	200	20	28.9	25
Viking	29	300	26.1	21	27
Provost	28	385	30	24	15
Suffield	26	n/a	32	14.23	10

Table 4.2: Real waterflooding projects across the US. Data reproduced from H. Saboorian-Jooybari, M. Dejam, and Z. Chen, Heavy oil polymer flooding from laboratory core floods to pilot tests and field applications: half-century studies, *J. Pet. Sci. Eng.*, 2016, 142, 85. Copyright: Elsevier (2016).

Eight cases of successful polymer flooding were reported across the world as shown in Table 4.3. The maximum oil recovery and optimum polymer flooding (34%) were obtained in the Marmul oilfield, located in the southern part of Oman. The main reasons behind the high percentage of oil recovery in the Marmul oilfield were high rock permeability, high displaced (oil) viscosity, low saline formation water, and moderate reservoir oil which resulted in sharp oil recovery.

This polymer injection project was the first pilot polymer-flooding test, which was undertaken in the Middle East (1986).

Oilfields	Injected pore volume (PV) size (bbl)	Oil recovery (%)
Westcat canyon	2000	15-20
Marmul	0.63	34
Bohai bay	0.3	7.6
Pelican lake	0.35	10-25
East bodo	n/a	9
Mooney	0.54	12
Suffield	0.6	10
Steal	0.1	5-6

Table 4.3: Successful heavy oil polymer flooding across the US. Data reproduced from H. Saboorian-Jooybari, M. Dejam, and Z. Chen, Heavy oil polymer flooding from laboratory core floods to pilot tests and field applications: half-century studies, *J. Pet. Sci. Eng.*, 2016, 142, 85. Copyright: Elsevier (2016).

A top-down understanding of successful polymer flooding has been illustrated in Figure 4.13. All EOR companies have agreed the steps of this process. Initially, preliminary results (experimental and modeling) for various polymers were used to assess and enhance the efficiency of flooding in the simulated reservoir (in HPHT condition). Two issues at the forefront of polymer flooding are that of cost and effective viscosity. By selecting the right polymer, according to the reservoir conditions, these issues could be addressed, and oil recovery significantly improved. Upon completion of successful simulation experiments, pilot and field test experiments can be performed.

Demulsifiction of water/oil in oil reserves

In the oilfields, emulsification of oil and water is truly a dilemma. Natural oil components include resin, asphaltenes, wax, and carboxylic acids, which perform as a surfactant, also making the oil-water mixture stabilize (Zhang et al., 2015). The best way to get rid of emulsification is by using demulsifiers to break the emulsion phase before their transportation in the pipeline and refining process. Chem

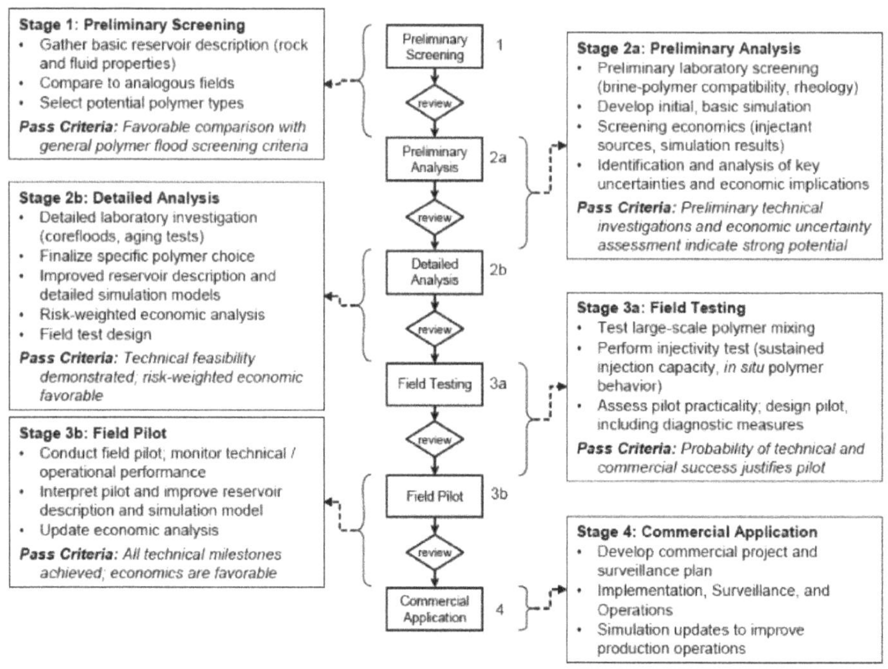

Figure 4.13: Steps of a successful polymer-flooding scenario across the oil reservoirs. Reproduced with permission from H. Saboorian-Jooybari, M. Dejam, and Z. Chen, Heavy oil polymer flooding from laboratory core floods to pilot tests and field applications: half-century studies, *J. Pet. Sci. Eng.*, 2016, 142, 85. Copyright: Elsevier (2016).

ical demulsifiers are amphiphilic compounds (both hydrophilic and hydrophobic groups), which have been widely utilized for IFT reduction and phase separation. Some polymeric surfactants are demulsifiers (Zhang et al., 2015; Pacheco et al., 2011). They can adsorb at the oil-water interface and make uniform oil front in the rock formation. Nonionic, cationic and anionic demulsifiers can be employed in the reserves. Nonionic emulsifiers have been reported for their high stability in the harsh environments. Ethylcellulose (EC, Figure 4.14), branched poly(ethylene oxide)−poly(propylene oxide) (PEO−PPO) block copolymers, branched PEO−PPO copolymer with hydrophilic segments (EO and OH), polyethyleneimine (PEI, Figure 4.15) with various ethylene oxide (EO)-propylene oxide (PO) unit, and polyamidoamine (PAMAM, Figure 4.16) are commercial demulsifiers to break water-in-diluted bitumen. Similar to other demulsifiers, hyperbranched polyglycerol (HPG, Figure 4.17) derivate is designed as a demulsifier with an average oil droplet size of <2 µm (Figure 4.18) (Zhang et al., 2015; Kainthan et al., 2006; Sunder et al., 1999).

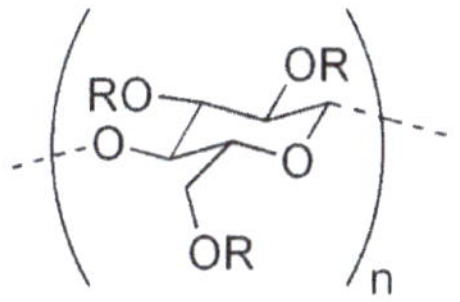

Figure 4.14: Structure of ethyl cellulose, which is a derivative of cellulose in which some of the hydroxyl groups (R = H) on the repeating glucose units are converted into ethyl ether groups (R = C_2H_5).

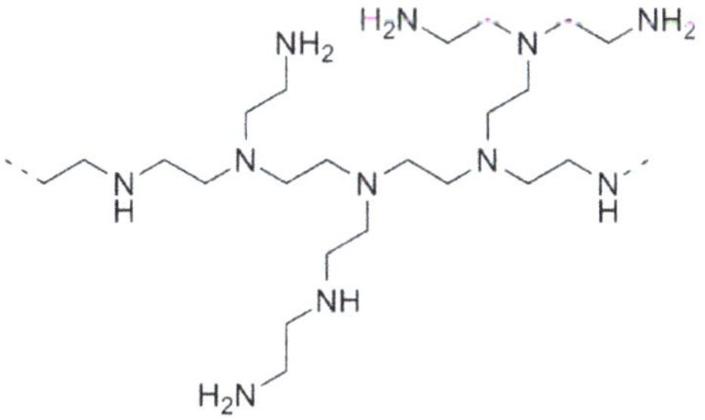

Figure 4.15: Structure of branched polyethylenimine (PEI).

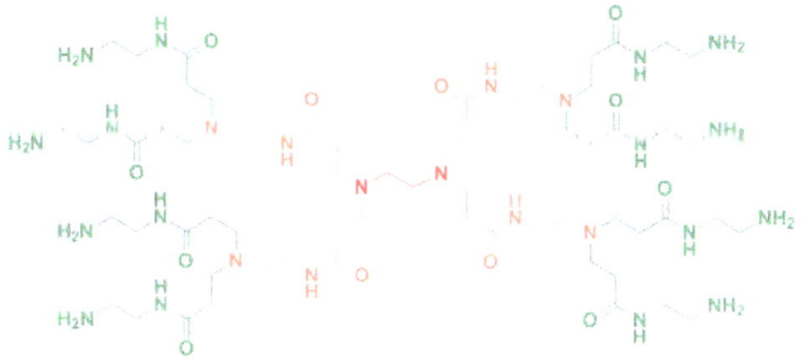

Figure 4.16: Structure of a generation 2 poly(amidoamine) (PAMAM) dendrimer.

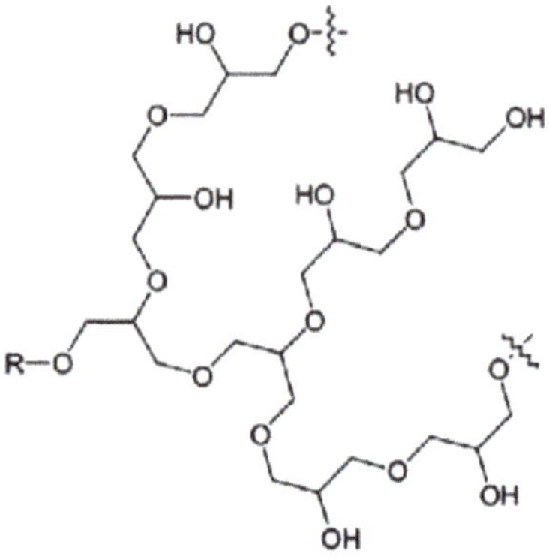

Figure 4.17: Structure of hyperbranched polyglycerol (HPG).

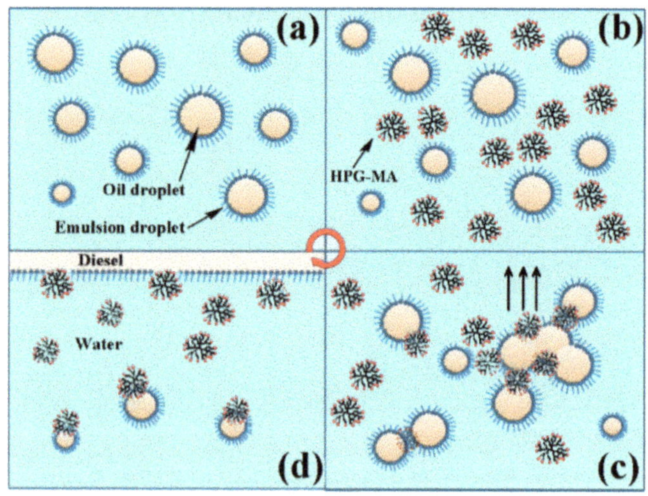

Figure 4.18: Schematic representation of the demulsification process used in oil reservoirs saturated by asphaltene aggregates. Asphaltene promotes coalescence of oil droplets, which leads to pore plugging through the rock matrix. The addition of demulsifier results in the separation of large to small oil droplets and consequently, increased oil recovery. The demulsification process: (a) O/W emulsion, (b) adding HPG-MA as a demulsifier, (c) aggregation and coalescence, and (d) separation. Reproduced with permission from L. Zhang, G. He, D. Ye, N. Zhan, Y. Guo, and W. Fang,). Methacrylated hyperbranched polyglycerol as a high-efficiency demulsifier for oil-in-water emulsions, *Energy Fuels*, 2016, 30, 9939.

Ethylene oxide-propylene oxide (EO-PO, Figure 4.19) compound is one of the main demulsifiers which causes an increase in the film-thinning rate and reduces the stability of film that forms in oil-water emulsions. EO- hydrophilic and -PO hydrophobic units generate an amphiphilic property in the emulsion and break the interfacial layer whose surface activity is superior to that of the highly polar component such as resin and asphaltenes (Wu et al., 2005). Other derivatives of EO/PO, such as polyoxyalkylated diethylene triamine EO/PO, PPO–PDMS–PPO blocks copolymer, and α-γ-diamines of polyoxyethylene–polyoxypropylene–polyoxyethylene (POE–POP–POE) triblock copolymer, were also shown to be suitable candidates for oil-water demulsification (Wang et al., 2010).

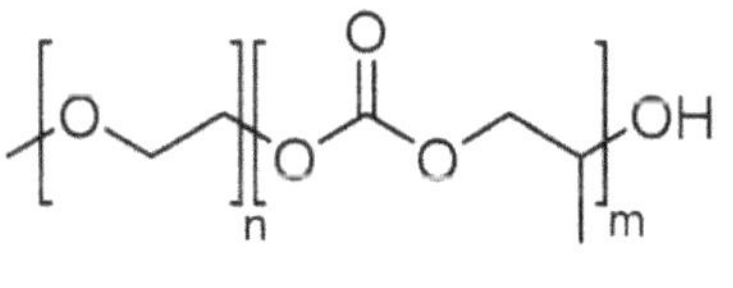

Figure 4.19: Structure of ethylene oxide-propylene oxide (EO-PO).

Bibliography

A. M. Firozjaii and H. R. Saghafi, Review on chemical enhanced oil recovery using polymer flooding: Fundamentals, experimental and numerical simulation, *Petroleum*, 2019, DOI:10.1016/j.petlm.2019.09.003.

P. D. Fletcher, L. D. Savory, F. Woods, A. Clarke, and A. M. Howe, Model study of enhanced oil recovery by flooding with aqueous surfactant solution and comparison with theory, *Langmuir*, 2015, **31**, 3076.

U. T. Gonzenbach, A. R. Studart, E. Tervoort, and L. J. Gauckler, Ultrastable particle-stabilized foams, *Angew. Chem., Int. Ed.*, 2006, **45**, 3526.

S. Gou, S. Li, M. Feng, Q. Zhang, Q. Pan, J. Wen, Y. Wu, and Q. Guo, Novel biodegradable graft-modified water-soluble copolymer using acrylamide and konjac glucomannan for enhanced oil recovery, *Ind. Eng. Chem. Res.*, 2017, **56**, 942.

R. K. Kainthan, E. B. Muliawan, S. G. Hatzikiriakos, and D. E. Brooks, Synthesis, characterization, and viscoelastic properties of high molecular weight hyperbranched polyglycerols, *Macromolecules*, 2006, **39**, 7708.

S. K. Kumar, N. Jouault, B. Benicewicz, and T. Neely, Nanocomposites with polymer grafted nanoparticles, *Macromolecules*, 2013, **46**, 3199.

V. F. Pacheco, L. Spinelli, E. F. Lucas, and C. R. Mansur, Destabilization of petroleum emulsions: evaluation of the influence of the solvent on additives, *Energy Fuels*, 2011, **25**, 1659.

H. Saboorian-Jooybari, M. Dejam, and Z. Chen, Heavy oil polymer flooding from laboratory core floods to pilot tests and field applications: Half-century studies, *J. Pet. Sci. Eng.*, 2016, **142**, 85.

T. Saigal, H. Dong, K. Matyjaszewski, and R. D. Tilton, Pickering emulsions stabilized by nanoparticles with thermally responsive grafted polymer brushes, *Langmuir*, 2010, **26**, 15200.

T. Saigal, A. Yoshikawa, D. Kloss, M. Kato, P. L. Golas, K. Matyjaszewski, and R. D. Tilton, Stable emulsions with thermally responsive microstructure and rheology using poly (ethylene oxide) star polymers as emulsifiers, *J. Colloid Interface Sci.*, 2013, **394**, 284.

R. Singh and K. K. Mohanty, Synergy between nanoparticles and surfactants in stabilizing foams for oil recovery, *Energy Fuels*, 2015, **29**, 467.

X. Su, M. F. Cunningham, and P. G. Jessop, Switchable viscosity triggered by CO 2 using smart worm-like micelles, *Chem. Commun.*, 2013, **49**, 2655.

A. Sunder, R. Hanselmann, H. Frey, and R. Mülhaupt, Controlled synthesis of hyperbranched polyglycerols by ring-opening multibranching polymerization, *Macromolecules*, 1999, **32**, 4240.

D. Wasan, A. Nikolov, and K. Kondiparty, The wetting and spreading of nanofluids on solids: Role of the structural disjoining pressure, *Curr. Opin. Colloid Interface Sci.*, 2011, **16**, 344.

J. Wu, Y. Xu, T. Dabros, and H. Hamza, Effect of EO and PO positions in nonionic surfactants on surfactant properties and demulsification performance, *Colloid Surf. A Physicochem. Eng. Asp,* , 2005, **252**, 79-85.

A. J. Worthen, S. L. Bryant, C. Huh, and K. P. Johnston, Carbon dioxide-in-water foams stabilized with nanoparticles and surfactant acting in synergy, *AIChE J.*, 2013, **59**, 3490.

L. Zhang, G. He, D. Ye, N. Zhan, Y. Guo, and W. Fang, Methacrylated hyperbranched polyglycerol as a high-efficiency demulsifier for oil-in-water emulsions, *Energy Fuels*, 2016, **30**, 9939.

J. Zhao, C. Dai, Q. Ding, M. Du, H. Feng, Z. Wei, A. Chen, and M. Zhao, The structure effect on the surface and interfacial properties of zwitterionic sulfobetaine surfactants for enhanced oil recovery, *RSC Adv.*, 2015, **5**, 13993.

Chapter 5: Nanofluid Enhanced Oil Recovery

Nanoparticles (NPs) are a unique class of materials with a size range of between 1-100 nanometer (nm). NPs are categorized according to their size, morphology, and chemical structures (Suleimanov et al., 2011). Based on physicochemical properties, they are classified into metal, carbon-based, semiconductor, ceramic, lipid-based, quantum dots (QDs), and polymeric NPs. Among these well-established NPs, metal-based and ceramic NPs exhibited the best physicochemical performance in EOR methods. Their derivatives have gained considerable interest in enhanced oil recovery (EOR) due to their extraordinary mechanical, chemical, and physical properties, ever since the first utilization of surface-modified nanoparticles in the year 2002. This chapter comprehensively reviews the flooding of metal and non-metal nanoparticles as a novel chemical EOR technique to ender more than 50% unrecoverable, 'trapped', oil after pushing primary and secondary oil recoveries. Moreover, the emerging advances regarding the use of these materials on EOR methods to overcome strong chemical bonding between rock-fluid and fluid-fluid interfaces are summarized.

Significant advancements of NPs in chemical EOR methods

In the last 10 years, various nanoparticles (metal and non-metal oxide) in the form of nanofluids, nano catalyst, and nano emulsion have been proposed for increased oil sweep efficiencies (Bera et al., 2012; Yoon et al., 2016). Their excellent properties such as particle size, shape, physiochemical stability, cost-effectiveness and eco-friendly processes have given fresh impetus to ongoing field-test applications for the future. In terms of nanoparticle application, they can be utilized as an emulsifier, catalyzer, and an effective agent in fluid for various EOR methods such as water, foam, polymer, surfactant, and thermal flooding (Guo et al., 2017; Son et al. 2016). Without a doubt, state-of-the-art additives such as nanomaterials along with polymers, and surfactants are extremely useful in maximizing oil recovery (Kmetz et al., Zhang et al.). Therefore, compared to secondary and tertiary recoveries, applying engineered nanoparticles has several advantages when sweeping inaccessible oil in reserves.

Their fascinating properties result in several key advantages in comparison to conventional EOR methods:
- They effectively reduce interfacial tension and improved contact angle while reducing interfacial tension, which can, for instance, provide excellent modification on rock wettability (oil-wet to water-wet);
- easily spontaneous oil displacement by changing gravity drainage between pores, pore-throats;
- increased sweep efficiency and a gravity reduction within the pores by diffusion mechanisms;
- increase the mobility ratio by reducing the oil viscosity;
- generating disjoint pressure for oil displacement through porous media (Wasan et al., 2011).

By considering the effect of NPs, we summarize the main effects on EOR flooding as demonstrated in recently published articles.

An example of wettability alteration

It has been reported that altering rock wettability can play a pivotal role in increasing oil production. Rock wettability alteration is a key strategy, consisting of controlling the relative permeability, residual oil distribution, and capillary pressure in pore and pore-throats. It has been stated that having a host rock with relatively water-wet and neutral-wet is the best situation in the oil reserves. Typically, polymer, surfactant, and alkaline-surfactant-polymer were designed within water flooding to improve spontaneous imbibition.

It has been claimed in colloidal materials that adding NPs leads to significant changes in interfacial tension and surface tension at fluid-fluid and rock-fluid interfaces due to changes in the entropy of NPs (Figure 5.1) (Pu et al., 2016). For instance, the large quantities of NPs lead to a wedge-film structure at the fluid (oil)/solid (rock) interface due to the physicochemical interaction by the NPs (Figure 5.2). Therefore, focusing on the surface-interface mechanisms of NPs resulting in the successful design of a NPs-chemical EOR operation.

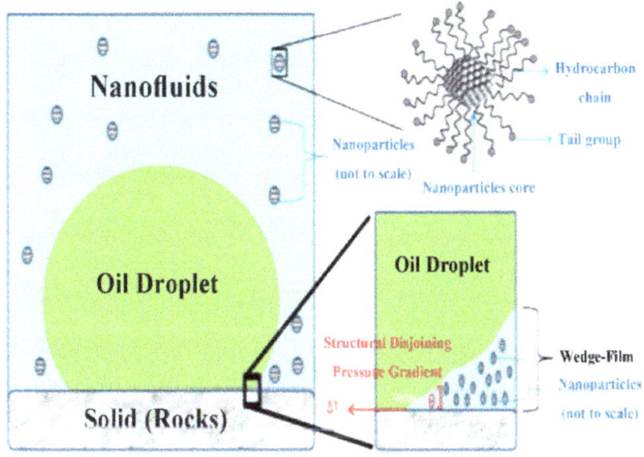

Figure 5.1: Schematic of oil-solid interaction mechanism in the presence of nanofluids, generating a wedge film via the diffusion of nanoparticles between rock-fluid interface. Reproduced with permission from K.-L. Liu, K. Kondiparty, A. D. Nikolov, and D. Wasan, Dynamic Spreading of Nanofluids on Solids Part II: Modelling, *Langmuir*, 2012, 28, 16274. Copyright: American Chemical Society (2014).

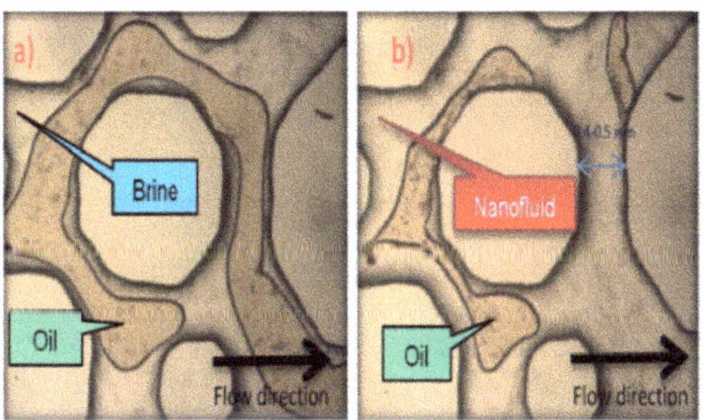

Figure 5.2: Nanoparticle sweeping mechanisms through the micromodel pores (a) after saline injection and (b) nanofluid injection. Reproduced with permission from K.-L. Liu, K. Kondiparty, A. D. Nikolov, and D. Wasan, Dynamic Spreading of Nanofluids on Solids Part II: Modeling, *Langmuir*, 2012, 28, 16274. Copyright: American Chemical Society (2014).

Examples of NPs adsorption on host rock

As the above content shows, changing the rock wettability with nanostructures is a topic of great interest for oil recovery. More examples have been shown to improve the understanding of this topic. Li examined the use of different hydrophilic and hydrophobic SiO_2 NPs by assessing the oil recovery values of Berea sandstone core plugs. They summarized the adsorption isotherms, spontaneous imbibition, contact angle measurements, and strengths and weakness of SiO_2 NPs.

Using explanatory data from two NPs with different functionalities, it was shown that hydrophilic SiO_2 makes the core relatively water-wet, whilst using hydrophobic SiO_2 does not cause a change in the wettability of origin rock (Roustaei & Bagherzadeh, 2015). They conclude that such hydrophilic NPs (with low contact angle, 10-20°) have a high tendency to adsorb on the rock surface, highlighting their use in the application within sandstone oil reservoirs. Using NPs to change the rock wettability of the carbonated reservoirs has also been strongly recommended by literature. For example, researchers analyzed Al_2O_3, TiO_2, and SiO_2 NPs and found that the rate of adsorption increased significantly when compared to systems without NPs (Ehtesabi et al., 2015; Kiani et al., 2016).

pH value

Applying nanomaterials in the oil reserves has several advantages and disadvantages compared with the conventional chemical substances. One of the major issues with NPs is their pH-responsive property in harsh reservoir media, which during the past 20 years have not been properly addressed. Changing the pH of NPs not only affects their aggregation in the pores and pore-throats but also owes to pore plugging in the oil formation. The surface of NPs using surfactant as well as adding functionalized groups are two useful approaches which can help researchers and engineers in resolving this issue. It has been reported that higher oil recovery can be made possible by optimization of the stability of NPs in the reservoir of varying pH, as evaluated by isoelectric point (IEP).[1] If the hydration repulsion between NPs in aqueous electrolyte suspension is high, the IEP value of a nanofluid is near zero and the particles tend to cluster and create an unstable fluid. While the optimum values will be obtained through high IEP values with ultra-stable nanofluids, it is noted that the lowest and highest stabilities within solutions are when ζ values are above or below 30 mV (±30 mV), respectively. pH profiles in different reservoirs with lower/higher concentration depend on the amount of monovalent and divalent cations such as Mg^{2+}, Ca^{2+}, and Cl^- in harsh conditions, in which the presence of ions can generally promote adsorption of NPs with ions on the substrate. Furthermore, shape, surface area, pore-volume, pore size, and NPs functionality are the main factors that cause increments in fluid stability.

An in-depth understanding of pH-responsive nanomaterial is required before implementing any chemical EOR flooding in the oil reserves. Most NPs are influenced by inorganic salts in an aqueous solution, and that they have a high tendency to agglomerate and deposit in the presence of different ions, which leads to both changes in IFT and wettability. Nonetheless, it is expected that the performance of NPs should be improved by surface functionalization during synthesizing or surface modification by surfactant, polymers, and ultrasonic vibration. By applying ultra-stable NPs, interfacial tension will decrease, and oil sweep efficiency will improve due to surface chemistry compatibility and stability on the bulk surface of NPs.

Successfully stabilized NPs with high thermal stability in electrolyte solutions were recently synthesized in foam and Pickering emulsion (Saigal et al., 2010; Capek, 2004; Chevalier & Bolzinger, 2013). One well-known instance is the stabilized anisotropic particles, which exhibits highly stabilized Pickering emulsion.[2] A significant change in reserved oil bank occurs when using surface-modified NPs during

---

[1] The isoelectric point (IEP) is the pH at which a molecule carries no net electrical charge or is electrically neutral in the statistical mean.

[2] A Pickering emulsion is an emulsion that is stabilized by solid particles (e.g., colloidal silica), which adsorb onto the interface between the two phases.

foam and Pickering emulsion flooding in the aqueous electrolyte solutions (Saigal et al., 2013). For instance, upon the surface modification of sphere silica NPs with polyethylene glycol (PEG, Figure 5.3), dramatic stabilized water/CO_2 emulsion and homogenous foam fronts were observed during Pickering emulsion and foam flooding, respectively.

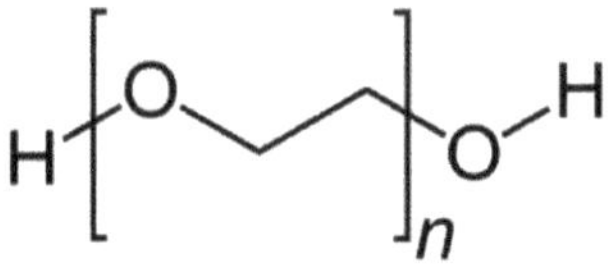

Figure 5.3: Structure of polyethylene glycol (PEG).

Much like silica particles, other NPs such as cellulose (a plant-based material, see Figure 5.4) also undergo significant stability within electrolyte solutions in EOR scenarios (Worthen et al., 2013). Surface modified nanocellulose (NC) have been evaluated in order to improve the viscous properties of nanofluids. NC is a material composed of nanosized cellulose fibrils with a high aspect ratio (width = 5–20 nm, length = micrometers). This approach opened several exciting avenues for utilizing green nanomaterials that have shown to be effective in decreasing FT to 10^{-1} mN/m. By changing the pH value, weak interactions amongst NPs and rock formation occurs. These weak interactions between NPs and mono/divalent ions (i.e., Na^+, Mg^{2+}, Ca^{2+}, Fe^{2+}, Cl^-, CO_3^{2-}, and SO_4^{2-}) lead to positive formation wettability alteration (oil-wet to water-wet) and further oil mobility.

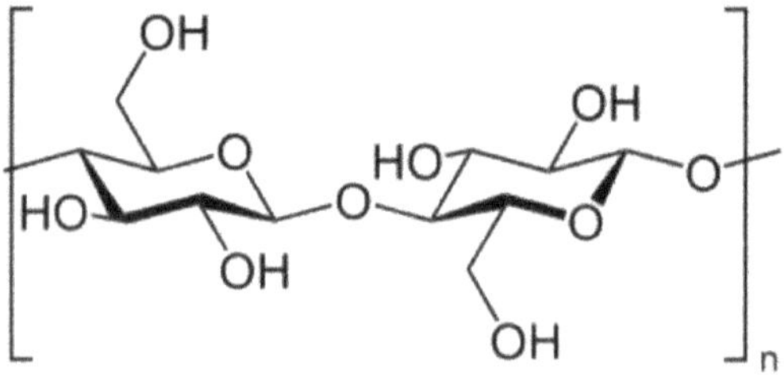

Figure 5.4: Structure of cellulose.

Effect of pH on deposition of clay NPs

Figure 5.5 shows the effect of pH on aggregation and stabilization of NPs. For instance, nano clay demonstrates great promise in changing the pH value; however, the chemistry of these layer-based NPs may not be tuned for different EOR scenarios.

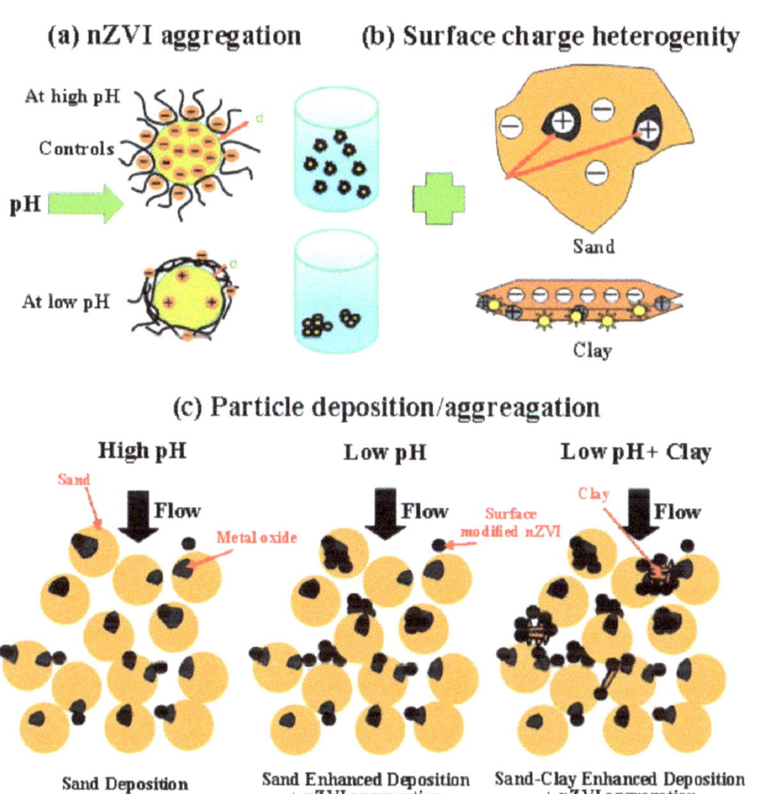

Figure 5.5: Schematic model of polymer aggregation onto surfaces in the presence of NPs onto different surfaces: (a) decreasing pH caused NPs aggregation, (b) at low pH, more deposition on positively charged edge sites on kaolinite or metal oxide impurities on sand, and (c) particle filtration by deposition of aggregates formed at low pH, and by deposition and mechanical filtration by clay aggregates at low pH. Reproduced with permission from H.-J, Kim, T. Phenrat, R. D. Tilton, and G. V. Lowry, Effect of kaolinite, silica fines and pH on transport of polymer-modified zero valent iron nanoparticles in heterogeneous porous media, _J. Colloid Interface Sci._, 2012, 370, 1. Copyright: Elsevier (2012).

The hydration process is the main issue in these NPs, which generates hydroxylation and protonation amongst cation and anion sites. Figure 5.5c shows the sensitivity of NPs to the positive charge (H^+ ions), which led to aggregation at low pH value; however, deposition of clay NPs was significantly reduced at higher pH value.

Recently, the effect of pH on long-term stability and aggregation of polyelectrolytes coated nano zero-valent iron (nZVI, Fe^0) onto kaolinite and sand surfaces has been investigated. It was observed that a decrease in NPs stability occurred with decreasing pH (Kim et al., 2010). Moreover, the presence of divalent ion concentration caused greater deposition onto clay minerals than silica surfaces. Another type of clay NP that was recently assessed as an emulsion droplet stabilizer in an electrolyte solution is montmorillonite $[(Na,Ca)_{0.33}(Al,Mg)_2(Si_4O_{10})(OH)_2 \cdot nH_2O]$. They proposed that each emulsion bubble was wrapped in multiple montmorillonite plate-sized NPs in order to stabilize them.

Impact of temperature on NPs

Chemical flooding continually deforms under high-pressure high-temperature (HPHT) condition and high salinity. These are the main issues for application of NPs, polymers, or surfactants in the oil formations. They should be considered before flooding to prevent phase segregation and molecular aggregation.

Impact of salinity on NPs

The main sources of salinity within reserves lie in rock and water formations, due to the accumulation of electrolytes formed over millions of years. The salinity of the rock formation has a significant effect on the success of any chemical floods; especially nanofluid. Therefore, applying a compatible geological test before implementing nanofluids would have an added benefit for successful chemical flooding. On the other hand, the reservoir's fluid chemistry is another challenging parameter. For instance, the presence of various divalent (Ca^{2+}, Mg^{2+}) and monovalent (K^+, Na^+) cations within nanofluid decreases the repulsive forces resulting in agglomeration of the NPs, which also occurs in polymer and surfactant molecules. Currently, there is no ultra-stable nanofluid system, so researchers are trying to improve fluid stability. Polyelectrolyte nanofluid system is available in the presence of KCl and NaCl to retain electrostatic repulsion forces between NPs and polymer in the high salinity mediums. As well as looking for a stabilized system, researchers found other existing polyelectrolyte systems such as polystyrene sulfonate (Figure 5.6a) and vinyl pyrrolidine (Figure 5.6b) can be used to help NPs stabilized.

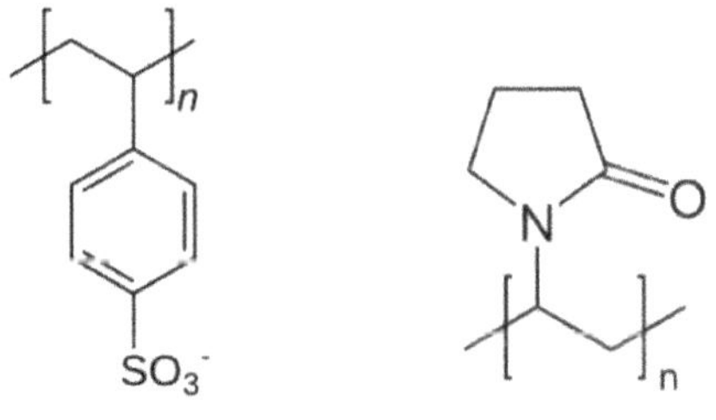

(a)   (b)

Figure 5.6: Structures of (a) polystyrene sulfonate and (b) poly-N-vinyl-2-pyrrolidine.

Effect of NPs surface modification

The particle-particle interaction and dispersion of nanoparticles in the base fluid is a great challenge in terms of applicability. Generally, two strategies have been suggested to dominate NPs dispersion in the fluid. One approach is covalent bonding using tune surface functional groups (ligands) such as silane, carboxylic, and amine. The second strategy is tailoring NPs to ligands such as surfactants, polymers, dendrimers,[1] biomolecule, and small molecules that generate a steric stabilization effect between the NPs. However, in acidic or basic solution NPs are unable to remain stabilized when lower steric

---

[1] Dendrimers are highly ordered, branched polymeric molecules. The name comes from the Greek word *Dendron* meaning *tree*. Typically, dendrimers are symmetric about the core, and often adopt a spherical three-dimensional morphology.

interactions are present. As opposed to neutral solutions, where this is not an issue and high NPs stability exists at lower steric interactions.

Systematic studies have shown that functionalized NPs can reach higher stability with ultra-low IFT, and contact angle to apply in foam and Pickering emulsion EOR methods (Chevalier & Bolzinger, 2013). A Pickering emulsion is an emulsion that is stabilized by solid particles (for example colloidal silica), which adsorb onto the interface between the two phases. Various metal and non-metal NPs such as silicon dioxide (SiO_2), zirconium dioxide (ZrO_2), titanium dioxide (TiO_2), magnesium oxide (MgO), aluminum oxide (Al_2O_3), copper dioxide (CuO), iron oxide (Fe_2O_3, Fe_3O_4), cerium dioxide (CeO_2), tin oxide (SnO_2), zinc oxide (ZnO), cobalt ferrite ($CoFe_2O_4$), and Janus NPs have been modified and used for oil recovery (Cheraghian et al., 2017; Ehtesabi et al., 2015; Jafarnezhad et al., 2017; Kiani et al., 2016).

Janus particles are special types of nanoparticles or microparticles whose surfaces have two or more distinct physical properties. For example, in order to synthesize a Janus zirconium phosphate NPs, the NPs surface was modified with octadecyl isocyanate and applied in EOR emulsion method. Janus zirconium phosphate NPs generates a fine emulsion with high oil recovery. In addition, adding alkyl poly-glycerol-side (APG) surfactant to zirconium phosphate nanofluid facilitate a positive synergistic effect, enhancing oil recovery and in turn, successful Pickering emulsion flooding (Sunder et al., 1999).

Controlling viscosity

Stable nanofluids are useful chemicals in reducing oil viscosity in heavy and extra-heavy oil reserves. One well-known example is ferrofluid, which showed excellent oil push up. Ferrofluid is composed of a nanoscale ferromagnetic fluid that becomes strongly magnetized via a magnetic field. Significant viscosity reduction of oil occurred by adding ferrofluid along with heat. Besides, ferrofluid leads to an increase in fluid viscosity displacement upon decreasing the oil viscosity. Both of these factors are the main driving forces behind the increase in mobility ratio and oil recovery. Another example of NPs demonstrating significant oil sweep efficiency is a SiO_2 NPs within a 300-ppm polymer solution with brine. It generated high oil mobilization within reservoirs due to the addition of NPs to the polymer solution, which can be attributed to varying monomer arrangement, intermolecular interactions, and "log-jammed" of SiO_2 at the pore throat. Besides SiO_2 NPs, other metal oxides like NiO_2 NPs within xanthan gum (Figure 5.7) solution and oleic acid were also used to improve viscosity and stability in the EOR process.

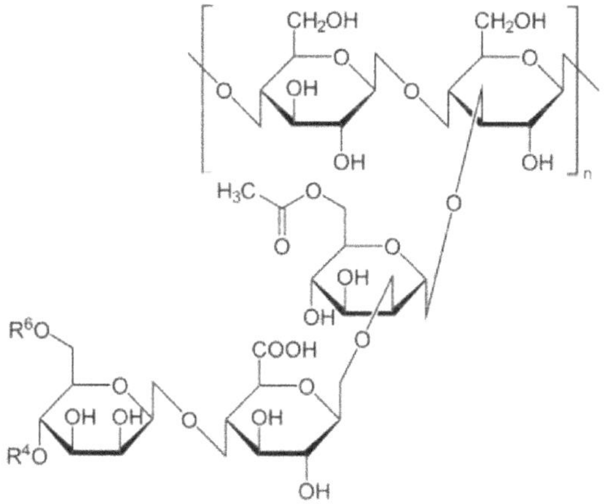

Figure 5.7: Structure of the polysaccharide xanthan gum.

Impact of NPs on asphaltenes adsorption

Over the past decade, heavy and extra-heavy oil, containing asphaltene (Figure 5.8) and resin, has been a major impetus for wide-ranging studies to decrease the strength of interactions between oil molecules and rock surfaces. With growing evidence of the effects of nanoparticles on increased oil sweeping efficiencies, some NPs can be used as an emerging platform in heavy oil (extra oil-wet) conditions. For instance, ZrO_2 nanofluids, when combined with non-ionic surfactant, result in significant oil recovery in carbonated heavy oil reserves. In this case, surface alteration from oil-wet to water-wet was observed; however, it was shown that wettability alteration requires at least two days, while the maximum rate of oil recovery occurred shortly after contact between the nanofluids and rock interface. A high affinity of rock, in order to change the rock surface to the super-hydrophilic condition, was also observed. Furthermore, ZrO_2-based nanofluid shows suitability as a wettability modifier that can result in considerable oil sweeping in fractured and low-permeability carbonate reservoirs. Utilizing SiO_2 in the presence of asphaltene aggregates is owed to their adsorption on the NPs surface. Adsorption of NPs on NPs surfaces has motivated research teams to work on novel NPs in this field.

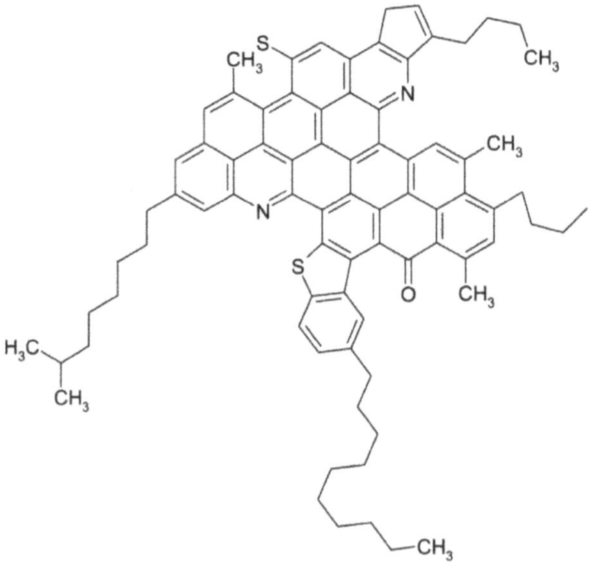

Figure 5.8: Proposed representative structure of asphaltene.

Coating highly polar asphaltene aggregates on SiO_2 NPs within the polymer is a way to improve the oil sweep efficiency in different pH values. For instance, adding NPs to long-chained polymer molecules in waterflooding leads to significant front displacement in the presence of asphaltene aggregates.

Nanoparticle classification

Metal-oxide NPs

Experimental studies have demonstrated the importance of metal NPs for EOR in foam and Pickering emulsion (Singh & Mohanty, 2015). They have been studied extensively during the past 20 years because NPs possess unique physicochemical properties. Metal oxide NPs such as Al_2O_3, MgO, and SiO_2 has a pH-responsive and magnetic-responsive property (Table 5.1). Due to these advanced properties, they can be used as a unique agent to modify the rock wettability, reduce oil viscosity, reduce fine migration, and improve oil sweep efficiency in the reserve. Free movement of metal NPs in the formation is crucial in the system. However, the mechanism of NPs movement is less considered. Figure 5.9 shows the feasible NPs movement amongst rock formation. NPs adsorption and retention are essential for easy movement of NPs under harsh conditions (high salinity, high pressure, and high temperature).

Nanoparticles	Oil type	Sand type	Main EOR mechanism	Oil recovery (%)
ZrO_2/water	Heavy oil	Carbonated	Aging time, type of nonionic surfactants	40%
SnO_2/brine	Heavy oil	Carbonated	NP type, concentration, and wettability alteration	39-61%
$CoFe_2O_4$	Medium oil	Sandstone	Reduce the oil viscosity	20-30%
Al_2O_3, SiO_2, NiO, TiO_2/brine	Heavy oil	Sandstone	NPs type	17-24%
SiO_2, Al_2O_3/brine	Medium crude oil	Sandstone	NP type and concentration	SiO_2 (9–14%) + Al_2O_3 (8–5%)
Fe_2O_3 + Al_2O_3 + SiO_2/brine	Mineral oil	Sandstone	Injection mode and salinity	1-10%
SiO_2/brine	Light oil	Sandstone	Permeability,	6-15%

			NP concentration and pore volume	
SiO$_2$/brine	Paraffinic oil	Sandstone	Time-length of flooding	8%
SiO$_2$/brine	Light oil	Micromodel	NP concentration, two phase flow behavior, emulsions and adsorption	N/A
SiO$_2$/brine	Light oil	Sandstone	NPs concentration	4-5.5%
SiO$_2$/brine	Light oil	Sandstone	NP type and concentration	5-15%
SiO$_2$/brine	Light oil	Micromodel	Ageing time	16-17%
SiO$_2$/biosurfactant/water	Heavy oil	Micromodel	NP concentration, shale orientation, length, distance, injection pressure	28-40%
SiO$_2$/brine	Heavy oil	Sandstone	NP concentration	9-26%
SiO$_2$, Al$_2$O$_3$,TiO$_2$/brine	Light oil	Sandstone	NPs type	7-11%
SiO$_2$/brine	Light oil	Micromodel	NP size, permeability, injection rate, rock wettability, temperature	~9%
SiO$_2$/brine	Mineral oil	Sandstone	Permeability and NP concentration	1-10%
SiO$_2$,NiO, Fe$_3$O$_4$/solution	Asphaltene + toluene	Micromodel	NP type and concentration	SiO$_2$ (23%), NiO (15%), Fe$_3$O$_4$(8%)
SiO$_2$/brine	Light oil	Sandstone	NP size and concentration	5-10%
SiO$_2$/brine	Mineral oil	Sandstone	NP size and	9-19%

			concentration	
SiO_2	Medium crude oil	Sandstone	Base fluid and NP type	13-24%
TiO_2/brine	Medium crude oil	Sandstone	NPs concentration	10-14%
CuO, NiO, Fe_2O_3/brine	Medium crude oil	Carbonated	NPs type	Fe_2O_3 (9%) + NiO (8%) + CuO (14%)
Fe_2O_3 + Al_2O_3 + SiO_2/brine	Mineral oil	Sandstone	Concentration	9-21%

Table 5.1: Summary of types of metal oxide NPs used in EOR flooding.

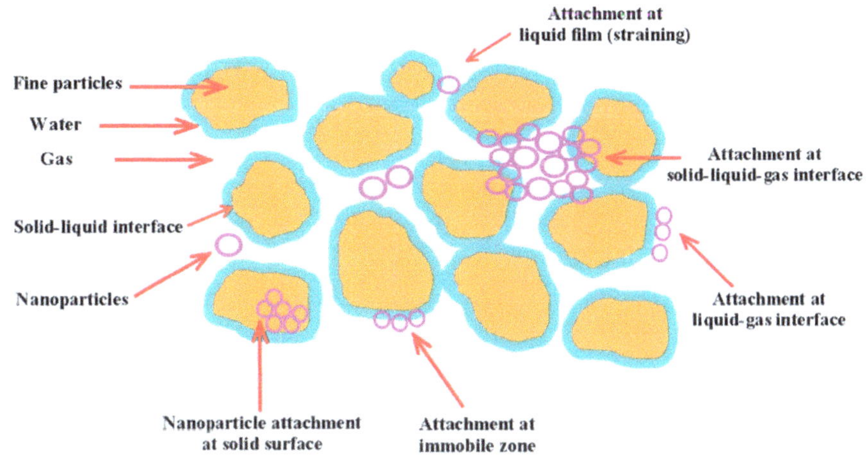

Figure 5.9: Schematic of NP interactions with rock through porous media.

Researchers have mainly focused on ceramic NPs like Al_2O_3 and SiO_2 to assess their potential properties after pressure decline in the reservoirs. Researcher's literatures reveal that the applicability of Al_2O_3 NPs in sandstone reservoirs are much greater than in carbonated formations. It has seen improved oil recovery upon adding 500 ppm Al_2O_3 NPs with surfactant solution that has led to changes in rock wettability from oil-wet to water-wet and increased the nanofluid viscosity in sandstone reservoirs. Furthermore, recently, engineers have compared the efficiency of Al_2O_3, titanium dioxide (TiO_2), and silicon dioxide (SiO_2) on limestone sample at 26, 40, 50, and 60 °C (Esfandiary Bayat et al., 2015). They have found 8.2%, 27.8%, 43.4% oil recovery for Al_2O_3, TiO_2, and SiO_2, respectively. They have confirmed the lowest and highest wettability alteration for Al_2O_3 and SiO_2 on the carbonated (limestone) formation. Besides, they found NPs can continue to change the contact angle for Al_2O_3, TiO_2, SiO_2 by 71±2°, 57±2°, and 26±2° on the limestone core plugs. Researchers expectations have risen sharply with the use of relatively unknown nanogel within Al_2O_3 solutions.[1] Nanogel is used with (Al_2O_3 NPs), to show signs of rheology improvement, which led to higher oil recovery in high permeability heterogeneous reservoirs.

Researchers estimated significant wettability alteration across the oil formation to be dependent on the NP properties such as functionality and concentration. For example, the silane is an efficient functionalized group used by scientists in order to stabilize nanofluid. It sticks on the NPs bulk surface featuring -OH group and triggers maximum NPs stability within the fluid. However, we will still require NPs stabilization to use them in the presence of brine for the foreseeable future. The polysilicon NPs (lipohydrophilic, hydrolipopholic and medium-wet NPs) will give this notice ahead of SiO_2 to be utilized in different EOR methods such as foam and emulsion (Singh & Mohanty, 2015).

Metal NPs in polymer solution

The recent addition of polymer to nanofluid marks a major step toward increased pseudoplasticity behavior in nanofluid. For example, adding partially hydrolyzed polyacrylamide (HPAM) to SiO_2 NPs improve both shear resistance and stability at high temperatures (Zhu et al., 2014). Researchers are working on the polymer with minimal sensitivity towards cost and loss for the oil reserves. However, synthesizing polymer with higher stability still proves challenging. Hopefully, many approaches such as the mixing of two di-block co-

---

[1] A *nanogel* is a nanoparticle composed of a *hydrogel*; a crosslinked hydrophilic polymer network.

polymers with hydrophobic monomers will help nanofluid to use polymers in the formation. Blending two block copolymers can help avoid aggregation and micellization, as the desired polymers for EOR application. By considering the aggregation number, monomer size, and response to external stimulus features, their applicability is closer in the reservoirs. Researchers have found for polymer to mix polyvinylpyrrolidone (Figure 5.10) with metal NPs (Al_2O_3, MgO, SiO_2, TiO_2) for high wettability alteration and oil recovery (11%). Moreover, oil saturation was reduced by 20% after addition of 600 ppm hydrolyzed polyacrylamide (HPAM) to SiO_2 NPs.

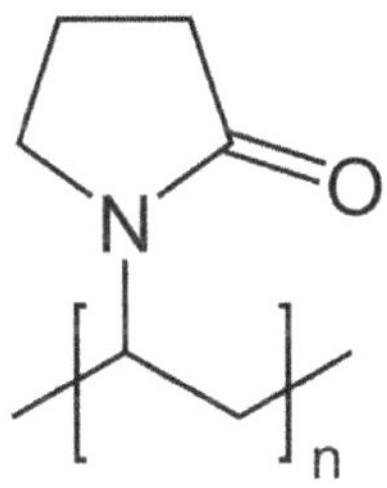

Figure 5.10: Structure of polyvinylpyrrolidone.

Furthermore, applying synthetic polymer such as polydimethylsiloxane (PDMS, Figure 5.11) with 1 wt.% CuO NPs in CO_2 injection makes a proper system with higher viscosity for heavy oil transport (API gravity 15) in sandstone reservoirs. The key observation made in this research is that using NPs (CuO-CO_2) with polymer rises the viscosity 140 times more than compared to just CO_2 injection. The final oil recovery reached 71% using polymer and CuO-CO_2 NPs.

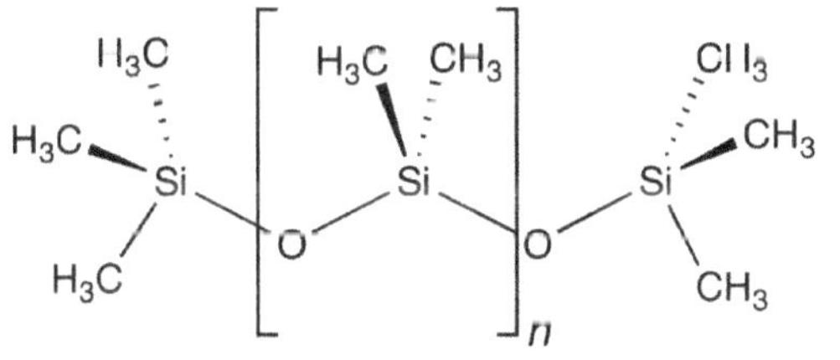

Figure 5.11: Structure of polydimethylsiloxane (PDMS).

Metal NPs in surfactant solution

A surfactant is an amphiphilic molecule composed of the hydrophilic head and hydrophobic tail. There is a range of surfactant types (both natural and synthetic) that can be applied, depending on the type of reservoir being treated, i.e., cationic, anionic, non-ionic, bio-surfactant, and zwitterionic. For EOR in sandstone reservoirs, anionic surfactants are preferred with varying head groups: sulfonate ($R\text{-}SO_3^-$), phosphate ($R\text{-}O\text{-}PO_3^{2-}$), sulfate ($R\text{-}O\text{-}SO_3^-$), and carboxylate ($R\text{-}CO_2^-$). The sulfonate group has been shown to improve the thermal stability of the surfactant at reservoir temperature (60 - 200 °C), as well as reducing the interfacial tension (IFT), altering the rock wettability, and having a lower tendency to adsorb on the surface of sandstone reservoirs. By contrast, phosphate and sulfated groups have been shown to give low IFTs and have better performance at lower temperatures, while carboxylated groups have been developed to improve the surfactant stability at high temperatures and high salinity (Karimi et al., 2016). Most metal NPs that have used with surfactant solutions are Al_2O_3, MgO, CuO, Fe_2O_3/Fe_3O_4, NiO, and Ni_2O_3 (Cortés et al., 2012). When in surfactant solutions, these metal NPs have a strong effect on the decrease in IFT and altering surface wettability in chemical EOR scenarios, such as the foam and emulsion. For example, very strong foam front and viscosity have been reported upon the addition of partially hydrophobic SiO_2 NPs to sodium dodecylbenzene sulfonate (SDBS, Figure 5.12).

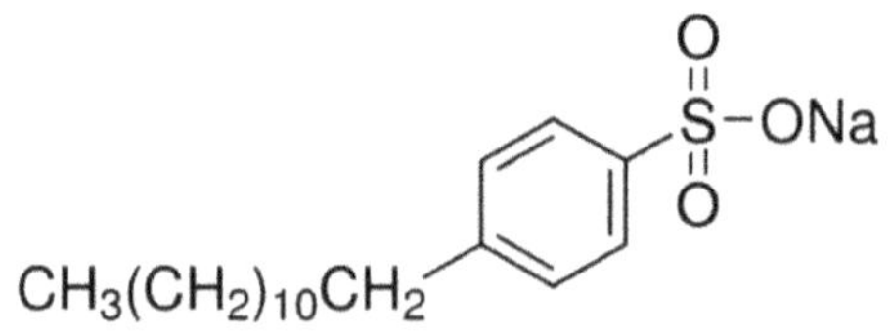

Figure 5.12: Structure of sodium dodecylbenzenesulfonate (SDBS).

Foam displacement with metal NPs

Foam is a dispersion of a high volume of gas (99 wt.%, N_2, CO_2, and air) into a small amount of liquid (1 wt.%). Gas purging and entrapment in surfactant solution leads to the adsorption of surfactant molecules into the gas-liquid interface, and as a result, this generates a continuous foam bubble with a thin-stable liquid/gas interface called lamellae. The effectiveness of foam in porous media is much more than water and gas. This leads to reduced gas relative permeability by bubble trapping and causes an increase in the gas viscosity, which provides stable and robust foam. The majority of reported cases have shown the foam system without any pore plugging. Figure 5.13 shows the possible mechanisms using hydrophilic nanoparticles arrangement at the interfaces by using a surfactant in foam generation (Su et al., 2013; Gonzenbach, et al., 2006).

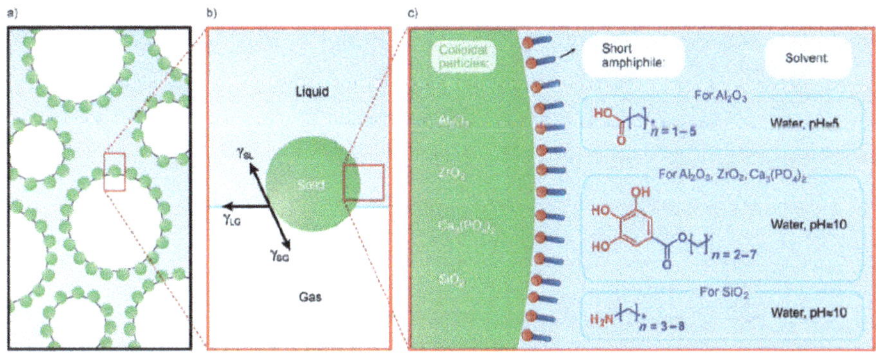

Figure 5.13: Schematic illustration of possible approaches in hydrophilic nanoparticles in foam generation: (a) stabilization of gas bubbles with colloidal particles; (b) the adsorption mechanism at the gas–liquid interface; (c) tuning the properties of various particles (such as Al_2O_3, ZrO_2, $Ca_3(PO_4)_2$, and SiO_2) by using surfactant to illustrate the foaming process. Reproduced with permission from U. T. Gonzenbach, A. R. Studart, E. Tervoort, and L. J. Gauckler, Ultrastable particle-stabilized foams, *Angew Chem.*, 2006, 45, 21. Copyright: Wiley-VCH (2006).

Carbon nanomaterials

Carbon nanostructures are made up of carbon atoms in a range size of 1 to 100 nm. Most carbon-based nanostructures have a large surface area to volume ratios compared with metal NPs. Carbon nanotubes (CNTs,), carbon-onion, graphene oxide, activated carbon, fullerene, and carbon black are the main types of carbon nanostructures (Kim et al., 2010). The properties of each nanocarbon are different due to the molecular interactions between carbon atoms. Nano carbons are applied in various sectors including electronic, biomedicine, catalysis, cosmetics, sensor, gas separation, and oil/gas industry. Nano carbons possess not only high surface areas but considerable surface functionality as well, which can be used as an effective foam and emulsion agent in EOR. A carbon nanotube has a unique superoleophilicity and can be used as an emulsion stabilizer in oil/water emulsion application. Furthermore, scientists discovered that, while the carbon nanotube (CNTs) grafts on graphene aerogel surface, the emulsion shows much more superhydrophobicity and oleophobicity.[1]

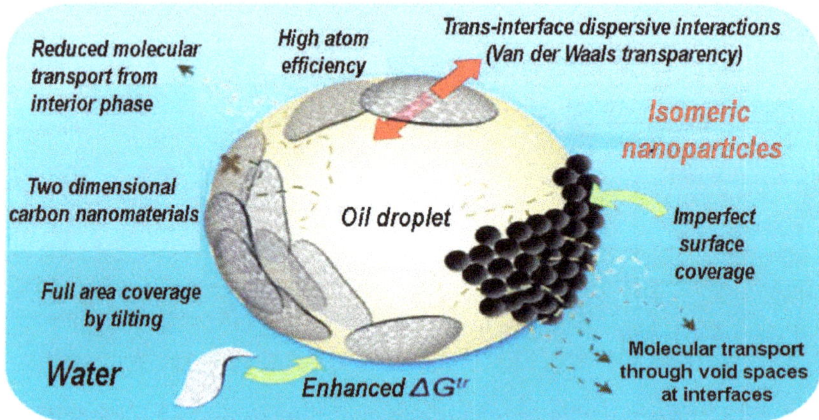

Figure 5.14: Schematic shows the mechanism of oil stabilization using graphene nanosheets and carbon black NPs over oil droplet. Reproduced with permission from M. A. Creighton, Y. Ohata, J. Miyawaki, A. Bose, and R. H. Hurt, Two-dimensional materials as emulsion stabilizers: interfacial thermodynamics and molecular barrier properties, *Langmuir*, 2014, 30, 3687. Copyright: American Chemical Society (2014).

The key feature in carbon nanotubes seems to be their ability to "tunning" the pore size and pore channel. Previous research has shown that a hybrid of silica NPs and carbon nanotube had a beneficial im-

---

[1] Oleophobicity describes how oils interact with a surface, and tends to imply that a surface rejects oils from it.

pact on the stabilization of emulsion by controlling the concentration and type of NPs (Li et al., 2016). Graphene, a two-dimensional nano-material, is another type of carbon nanomaterial (Figure 5.15) that can be utilized in EOR as an emulsion stabilizer (Figure 5.16).

Recently, modified graphene oxide (Figure 5.17) increased oil recovery ratio by 15.2%, via a strong elasticity force created by graphene, resulting in changes in the oil/water interactions at the interface. Moreover, when the researchers applied the amphiphilic graphene sheet to the emulsion, they found that dissociation of the carboxylic group due to changes in the pH value resulted in a tunable amphiphilicity.

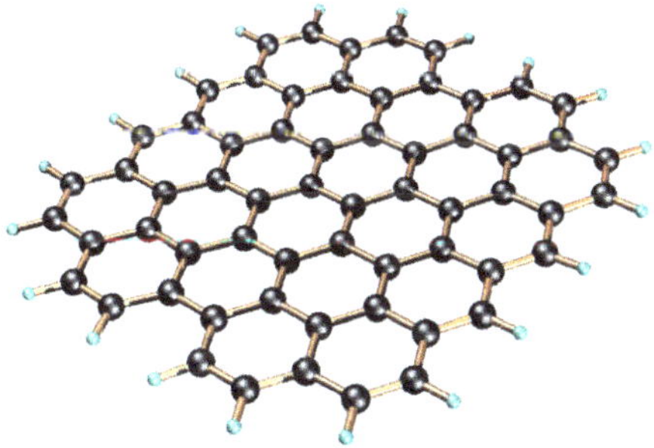

Figure 5.15: Idealized structure of a single graphene sheet.

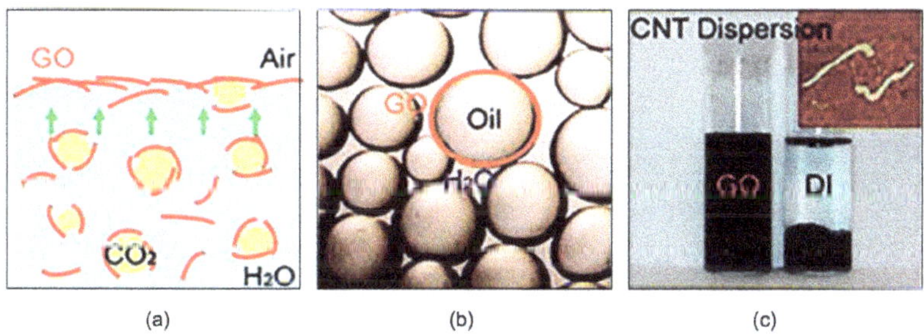

Figure 5.16: Graphene oxide at (a) gas-fluid, (b) fluid-fluid, and (c) solid-fluid interfaces. Reproduced with permission from J. Kim, L. J. Cote, F. Kim, W. Yuan, K. R. Shull and J. Huang, Graphene oxide sheets at interfaces, *J. Am. Chem. Soc.*, 2010, 132, 23. Copyright: American Chemical Society (2010).

Carbon nanomaterials in an emulsion, however, are predominantly particle geometry-based which results in changes in the unique properties of emulsion such as minimizing interfacial tension in foam and emulsion. Therefore, for a successful foam or emulsion EOR flooding, consideration of particle size and emulsion size is necessary, (under harsh reservoir conditions). However, highly stabilized and efficient emulsion as they may appear in the presence of CNTs or graphene oxide, increases the cost of oil production and new synthesis procedures should be taken into account. In other research, a highly stabilized emulsion system was generated using 300 ppm graphene oxide (GO)/polyacrylamide suspensions over the course of 28 days. Furthermore, to increase oil recovery, $CNTs/SiO_2$ was flooded on carbonated and sandstone reservoirs, which led to a reduction in IFT value and improvement in rock wettability.

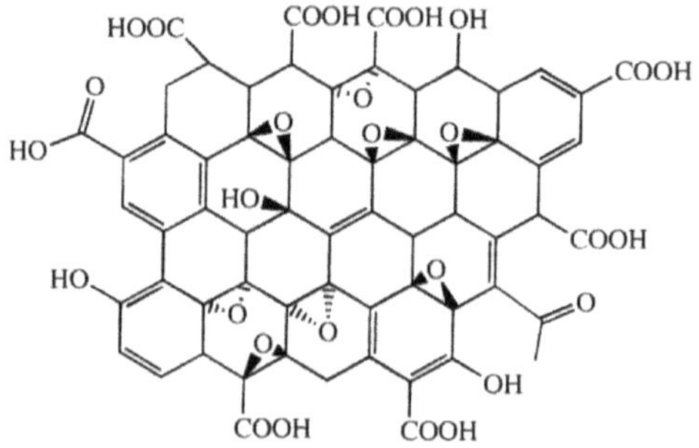

Figure 5.17: Idealized structure proposed for graphene oxide (GO). Adapted from C. E. Hamilton, PhD Thesis, Rice University (2009).

Carbon black is another form of carbon nanomaterial that is mainly applied in paints, inks, and plastics industries. It has been reported that the modified surface of carbon black can lead to an improvement in the fluid rheology and stabilize fluid in the reserves. Figure 5.18 shows the structure of ethylenediamine and acrylamide. The surface of CB covalently modified with ethylenediamine (EDA, Figure 5.18a) and acrylamide (AM, Figure 5.18b) is used to make stable CB-EDA-AM complexes for harsh condition.

(a)   (b)

Figure 5.18: Structures of (a) ethylenediamine (EDA) and (b) acrylamide (AM).

Quantum dot (QD) NPs

One of the most popular nanoparticles, sized between 2 to 10 nm (10 to 50 atoms in diameter) are quantum dots (QDs, Figure 5.19). They consist of a rod, core, core/shell and cadmium-free QDs (Haruna et al., 2019). The main method used to assess their size is via photon energy. Their small size allows them to become suspended within a fluid, which may prove useful in different applications such as EOR process.

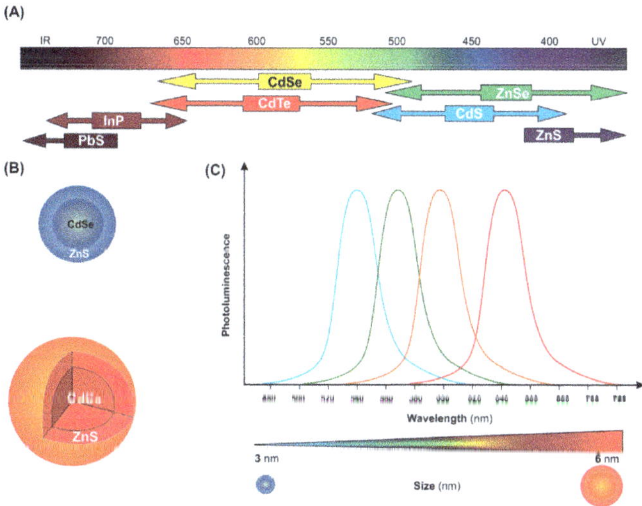

Figure 5.19: Different types of QDs Larger QDs. The particle with 2–3 nm diameter emits short wavelengths (blue and green color), while QDs of 5–6 nm diameter emits longer wavelengths (orange or red). Reproduced with permission from Y. R. Saadat, N. Saeidi, S. Z. Vahed, A. Barzegari, and J. Barar,). An update to DNA ladder assay for apoptosis detection, *BioImpacts*, 2015, 5, 1. Copyright: Tabriz University of Medical Sciences (2015).

Haruna et al. demonstrated that by adding QDs to HPAM solution, the fluid rheology decreases. However, they make good co-agents, being compatible with other substances in fluid transport. The main disadvantage is their sensitivity to brine solution that is yet to be addressed. Some strategies have been proposed to stabilize QDs into solutions. For instance, adding oleic acid to cadmium selenide (CdSe), which leads to higher fluid stability (0.55 M ionic strength and 1 M NaCl). In other research, stabilized carbon QDs were used in carbonated reservoirs in harsh (120,000 ppm TDS) and HPHT conditions (100 °C). However, at the time of writing this, it appears very unlikely that those QDs will apply, due to formation jamming and damage by NPs aggregation in real reservoir conditions (Haruna et al., 2019).

Oil tracer technology

Following earlier assessments, we deduced that a sustained decrease in conventional oil production is becoming increasingly likely before 2030. This information was provided without considering tight oil reservoirs due to them being classified as unconventional oil resources (He et al., 2015). By including tight oil reservoirs, the portion of oil banks drastically changes these results. Researchers and engineers are still working on being able to assess oil traces in the formations using advanced technologies. A large portion of these reserves can be explored and recovered by available technologies. Although, we have had to use other technologies to in reservoirs that are hard to be assessed by in the presence of oil.

Applying tracer technology using fluorescent nanoparticle is likely to trigger much exploration in tight and small oilfields. Tracing initiated by evaluation of real-time monitoring on the connectivity of the pores. Previous research has shown that fluorescent nanoparticles can be used during well drilling in the reservoir (within the mud system). Observation of this process is done via a wireline formation tester (WFT) string.[1] A fluorescent sensor then detects, records, and inter-

---

[1] A tool primarily to obtain formation pressures at chosen locations in an interval, and, with an accurate quartz gauge, permeability estimates may be obtained.

prets the data of contaminated fluids by fluorescent nanoparticles. According to the excitation light energy of each fluid (oil or water), the fluorescence sensor can provide clear differentiation between the drilling fluids, crude oil, and formation water in the flowing fluid. Florescent detectors, pyrene tetrasulfonic acid (PTSA, Figure 5.20), fluorobenzoic acid (FBA, Figure 5.21), and methyl, ethyl, isopropyl, and n-propyl acetates (Figure 5.22) are common tracers used in hydrocarbon reservoirs. Furthermore, this novel technology is being applied by a group of researchers as a powerful sensor to detect the number of hazardous gases such as H_2S and CO_2 in the reserves (Cubillos et al., 2015).

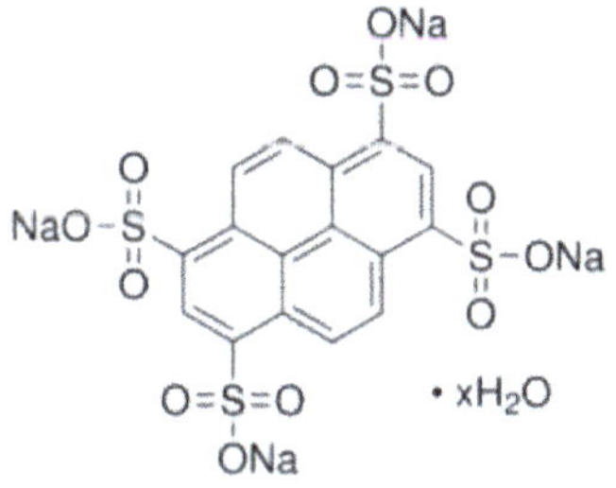

Figure 5.20: Structure of pyrene tetrasulfonic acid (PTSA).

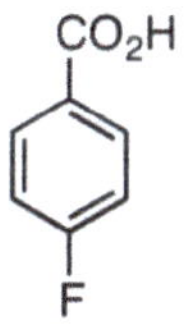

Figure 5.21: Structure of fluorobenzoic acid (FBA).

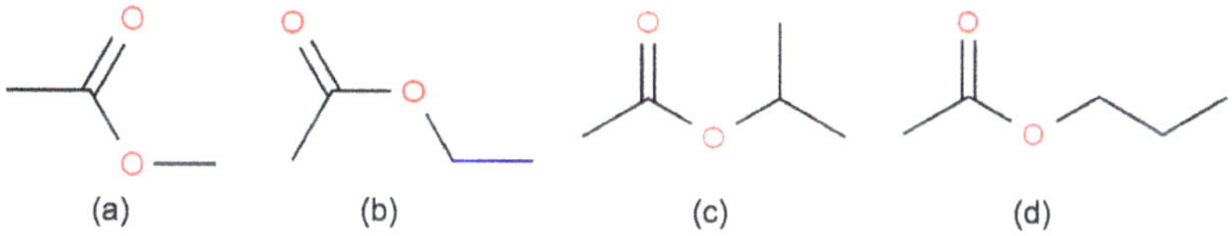

Figure 5.22: Structures of (a) methyl acetate, (b) ethyl acetate, (c) isopropyl acetate, and (d) n-propyl acetate.

There have been reports of Fe_2O_3, Fe_3O_4, and cobalt ferrite ($CoFe_2O_4$) NPs (smart ferrofluids) being utilized as a magnetic reservoir tracer (Cubillos et al., 2015). These types of NPs can reduce the amount of water cut and oil viscosity within steam-assisted gravity drainage (SAGD; "Sag-D"). SAGD is an EOR method used in heavy and extra heavy oil reservoirs.

Pickering emulsion

A Pickering emulsion consists of a fine dispersion of droplets following the agitation of two or more immiscible fluids. Pickering emulsion is type of emulsion with adsorbed solid particles over droplets (Pochert et al., 2016). Solid particles are within the range of 1-100 nm. A barrier layer is formed at the oil/water interface due to the presence of solid nanoparticles (Pickering emulsion) (Figure 5.23). Solid NPs are different types of solid minerals, synthetic polymers, polysaccharides, and proteins (Creighton et al., 2014). For instance, to retain oil stability from water, nanomaterials like SiO_2 NPs are added (Figure 5.23). They can be used in many applications such as food, cosmetics, biomedicine, and EOR. They are classified by oil-in-water (o/w), water-in-oil (w/o), and multiple phases. Their high resistance to emulsion droplet coalescence leads to improved rheology of displacing fluid in reservoirs. Their behavior is regulated by the size and morphology of solid particles over each droplet.

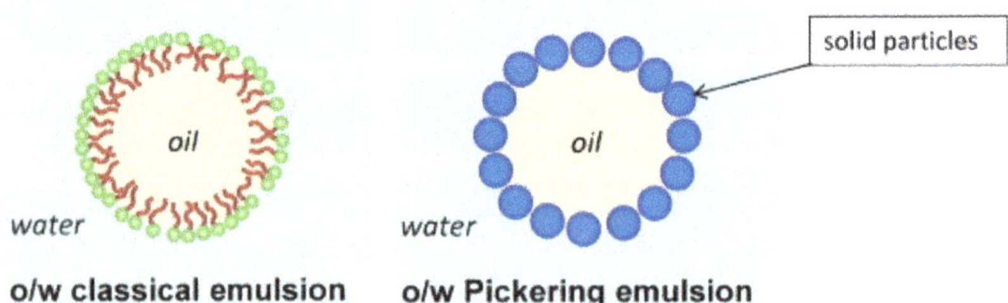

Figure 5.23: Comparison of a classical oil-in-water (o/w), and water-in-oil (w/o) Pickering emulsion. Reproduced with permission from Y. Chevalier and M.-A. Bolzinger, Emulsions stabilized with solid nanoparticles: Pickering, *Colloids Surf. A*, 2013, 439, 23. Copyright: Elsevier (2013).

Furthermore, many types of water-soluble polymers and surfactants such as polyacrylamide (PAM, Figure 5.24a), hydrolyzed polyacrylamide (HPAM), xanthan gum (Figure 5.7), polyacrylic acid (Figure 5.24b), cellulose (Figure 5.4), and starch (Figure 5.24c) along with some NPs such as (clay, SiO_2, CuO) were used under HPHT conditions. Adding polymers and surfactants to the Pickering emulsion, bringing about optimum rheology, higher stability, and controls sedimentation. In another example, researchers have succeeded in synthesizing a highly stabilized CNTs/SiO_2 nano-hybrid in harsh conditions.

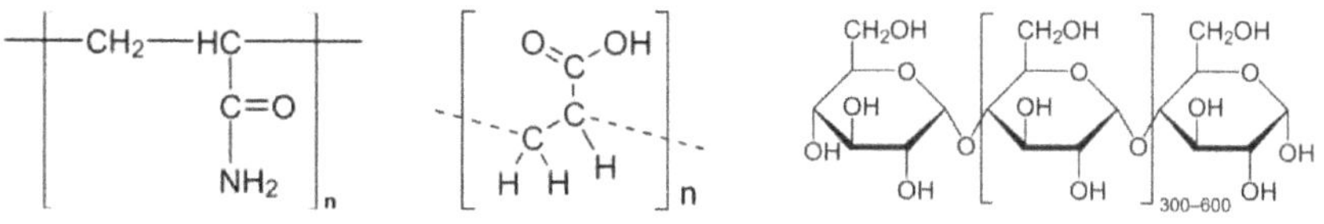

(a) (b) (c)

Figure 5.24: Structures of (a) polyacrylamide (PAM), (b) polyacrylic acid, and (c) starch.

Besides, the addition of surfactants such as hexadecyltrimethylammonium bromide (CTAB, Figure 5.25) and palmitic acid (Figure 5.26) to SiO_2 NPs generated a strong steric hindrance between NPs, consequently leading to higher emulsion stability.

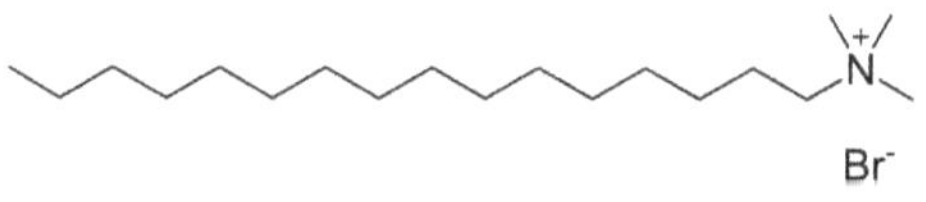

Figure 5.25: Structure of hexadecyltrimethylammonium bromide (CTAB).

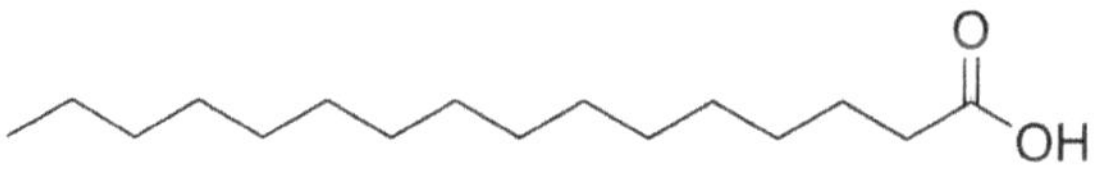

Figure 5.26: Structure of palmitic acid.

Incorporation of SiO_2 NPs, polystyrene, and polyacrylamide in Pickering emulsion resulted in strong adsorption of NPs and long-term stability at oil-water interface (Figure 5.27). On the other hand, demulsification in heavy oil reservoirs and bitumen extraction from oil sands has drawn a great deal of attention as a flexible approach to reduce oil viscosity and increase oil mobility (Haase et al., 2010; Kalashnikova et al., 2013; Liang et al., 2015).

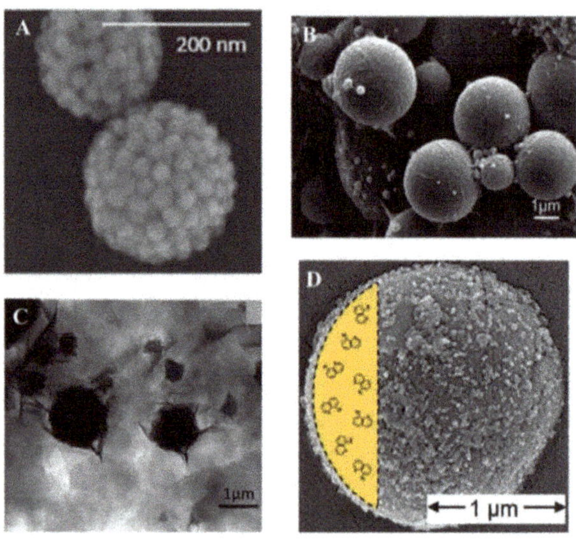

Figure 5.27: (a) the Cryo-SEM image of hydrophilic SiO_2 NPs stabilized by Pickering emulsions, (b) depicts how polystyrene particles coated by bacteria cellulose nanofibers, (c) shows polyacrylamide (PAAm) particles coated by functionalized montmorillonite, and (d) shows a pH-sensitive encapsulates covered by SiO_2 NPs. Adapted from:(a) S. Sihler, A. Schrade, Z. Cao, and U. Ziener, Inverse Pickering emulsions with droplet sizes below 500 nm, *Langmuir*, 2015, 31, 10392; (b) I. Kalashnikova, H. Bizot, P. Bertoncini, B. Cathala, and I. Capron, Cellulosic nanorods of various aspect ratios for oil in water Pickering emulsions, *Soft Matter*, 2013, 9, 952; (c) D. J. Voorn, W. Ming, and A. M. van Herk, Polymer–clay nanocomposite latex particles by inverse pickering emulsion polymerization stabilized with hydrophobic montmorillonite platelets, *Macromolecules*, 2006, 39, 2137; (d) M. F. Haase, D. Grigoriev, H. Moehwald, B. Tiersch, and D. G. Shchukin, Encapsulation of amphoteric substances in a pH-sensitive Pickering emulsion, *J. Phys. Chem. C*, 2010, 114, 17304.

Polymer-grafted nanoparticles (NPs)

Polymer grafted nanoparticles (PG-NPs) or hairy NPs (HNPs) is an integrated molecule, made from joining layers of polymer chains co-valently grafted on the NPs solid surface (Kumar et al., 2013). This approach may be used in different EOR methods such as emulsification, gas/water/polymer flooding, and foam injection (Li et al., 2019). The existing polymer has a profound effect on NPs stability in a harsh environment, which can result in higher stability in NPs via lower electrostatic forces between the polymer chains and high wettability alteration with lower adsorption on the rock, which is a suitable agent to sweep out trapped oil. NPs combination with surfactant is another useful approach similar to PG-NPs, which results in higher stability of NPs within a solution. Some PG-NPs decrease the IFT values by order of magnitude from 25 to 1 mN/m (Figure 5.28). For instance,

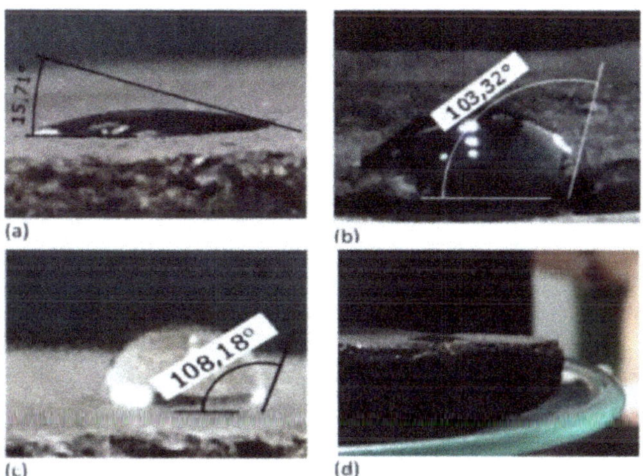

Figure 5.28: Photographic images showing the contact angle of (a) oil/air/rock before treatment, (b) oil/air/rock after treatment with silica NPs, (c) water/air/rock before treatment, and (d) water/air/rock after treatment with silica NP. Reproduced with permission from A. J. Worthen, S. L. Bryant, C. Huh, and K. P. Johnston, Carbon dioxide-in-water foams stabilized with nanoparticles and surfactant acting in synergy, *AIChE J.*, 2013, 59, 3490. Copyright: AICHE (2013).

grafting silica nanoparticles with polyacrylamide led to the higher viscosity and improving thermal stability in comparison with HPAM sharp under high temperature high salinity (HTHS condition; salinity: 32868 mg/L, at 85 °C) (Kumar et al., 2013). Earlier studies also confirmed grafted low chain polymer-NPs (0.1 wt.%) led to an increase in the system entropy and a decrease in its adsorptions on the rock in comparison with long-chain polymer-NPs.

Bibliography

B. P. Binks and A. Rocher, Stabilisation of liquid–air surfaces by particles of low surface energy, *J. Colloid Interface Sci.*, 2009, **335**, 94.

T. Capek, Preparation of metal nanoparticles in water-in-oil (w/o) microemulsions, *Adv. Colloid Interface Sci.*, 2004, **110**, 49.

G. Cheraghian, S. Kiani, N. N. Nassar, S. Alexander, and A. R. Barron, Silica nanoparticle enhancement in the efficiency of surfactant flooding of heavy oil in a glass micromodel, *Ind. Eng. Chem. Res.*, 2017, **56**, 8528.

Y. Chevalier and M.-A. Bolzinger, Emulsions stabilized with solid nanoparticles: Pickering emulsions, *Colloids Surf. A*, 2013, **439**, 23.

F. B. Cortés, J. M. Mejía, M. A. Ruiz, P. Benjumea, and D. B. Riffel, Sorption of asphaltenes onto nanoparticles of nickel oxide supported on nanoparticulated silica gel, *Energy Fuels*, 2012, **26**, 1725.

H. Cubillos, E. Yuste, M. Bozorgzadeh, J. Montes, H. Mayorga, S. Bonilla, G. Quintanilla, P. Lezana, A. Panadero, and P. Romero, The Value of Inter-well and Single Well Tracer Technology for De-Risking and Optimizing a CEOR Process-Caracara Field Case, 2015, SPE-174397-MS.

M. A. Creighton, Y. Ohata, J. Miyawaki, A. Bose, and R. H. Hurt, Two-dimensional materials as emulsion stabilizers: interfacial thermodynamics and molecular barrier properties, *Langmuir*, 2014, **30**, 3687.

H. Ehtesabi, M. M. Ahadian, and V. Taghikhani, Enhanced heavy oil recovery using TiO_2 nanoparticles: investigation of deposition during transport in core plug, *Energy Fuels*, 2014, **29**, 1.

A. E. Bayat, R. Junin, A. Samsuri, A. Piroozian, and H. Hokmabadi,). Impact of metal oxide nanoparticles on enhanced oil recovery from limestone media at several temperatures, *Energy Fuels*, 2014, **28**, 6255.

U. T. Gonzenbach, A. R. Studart, E. Tervoort, and L. J. Gauckler, Ultrastable particle-stabilized foams, *Angew. Chem.*, 2006, **45**, 21.

F. Guo and S. Aryana, An experimental investigation of nanoparticle-stabilized CO_2 foam used in enhanced oil recovery, *Fuel*, 2016, **186**, 430.

M. F. Haase, D. Grigoriev, H. Moehwald, B. Tiersch, and D. G. Shchukin, Encapsulation of amphoteric substances in a pH-sensitive Pickering emulsion, *J. Phys. Chem. C*, 2010, **114**, 17304.

Y. Hamedi-Shokrlu and T. Babadagli, Stabilization of nanometal catalysts and their interaction with oleic phase in porous media during enhanced oil recovery, *Ind. Eng. Chem. Res.*, 2014, **53**, 8464.

C. E. Hamilton, PhD Thesis, Rice University (2009).

M. A. Haruna, Z. Hu, H. Gao, H. J. Gardy, J. S. M. Magami, and D. Wen, Influence of carbon quantum dots on the viscosity reduction of polyacrylamide solution, *Fuel*, 2019, **248**, 205.

L. He, F. Lin, X. Li, H. Sui, and Z. Xu, Interfacial sciences in unconventional petroleum production: from fundamentals to applications, *Chem. Soc. Rev.*, 2015, **44**, 5446.

N. Hosseinpour, A. A. Khodadadi, A. Bahramian, and Y. Mortazavi, Asphaltene adsorption onto acidic/basic metal oxide nanoparticles toward in situ upgrading of reservoir oils by nanotechnology, *Langmuir*, 2013, **29**, 14135.

M. Jafarnezhad, M. S. Giri, and M. Alizadeh, Impact of SnO_2 nanoparticles on enhanced oil recovery from carbonate media, *Energ. Source. Part A*, 2017, **39**, 121.

I. Kalashnikova, H. Bizot, P. Bertoncini, B. Cathala, and I. Capron, Cellulosic nanorods of various aspect ratios for oil in water Pickering emulsions, *Soft Matter*, 2013, **9**, 952.

M. Karimi, R. S. Al-Maamari, S. Ayatollahi, and S. Mehranbod, Impact of sulfate ions on wettability alteration of oil-wet calcite in the absence and presence of cationic surfactant, *Energy Fuels*, 2016, **30**, 819.

S. Kiani, M. M. Zadeh, S. Khodabakhshi, A. Rashidi, and J. Moghadasi, Newly prepared Nano gamma alumina and its application in enhanced oil recovery: an approach to low-salinity waterflooding, *Energy Fuels*, 2016, **30**, 3791.

S. Kiani, A. Samimi, S. Khodabakhshi, and A. Rashidi, Novel one-pot dry method for large-scale production of nano γ-Al_2O_3 from gibbsite under dry conditions, *Monatsh. für Chemie-Chem.*, 2016, **147**, 1153-1159.

H.-J. Kim, T. Phenrat, R. D. Tilton, and G. V. Lowry, Effect of kaolinite, silica fines and pH on transport of polymer-modified zero valent iron nano-particles in heterogeneous porous media, *J. Colloid Interface Sci.*, 2012, **370**, 1.

J. Kim, L. J. Cote, F. Kim, W. Yuan, K. R. Shull and J. Huang, Graphene oxide sheets at interfaces, *J. Am. Chem. Soc.*, 2010, **132**, 23, 8180-8186.

A. A. Kmetz, M. D. Becker, B. A. Lyon, E. Foster, Z. Xue, K. P. Johnston, L. M. Abriola, and K. D. Pennell, Improved mobility of magnetite nanoparticles at high salinity with polymers and surfactants, *Energy Fuels*, 2016, **30**, 1915.

K. Kondiparty, A. Nikolov, S. Wu, and D. Wasan, Dynamic spreading of nanofluids on solids. Part I: experimental, *Langmuir*, 2011, **27**, 3324.

S. K. Kumar, N. Jouault, B. Benicewicz, and T. Neely, Nanocomposites with polymer grafted nanoparticles, *Macromolecules*, 2013, **46**, 319.

J. Liang, N. Du, S. Song, and W. Hou, Magnetic demulsification of diluted crude oil-in-water nanoemulsions using oleic acid-coated magnetite nanoparticles, *Colloids Surf. A*, 2015, **466**, 197.

X. Li, Y. Xue, M. Zou, D. Zhang, A. Cao, and H. Duan, Direct oil recovery from saturated carbon nanotube sponges, *ACS Appl. Mater. Interfaces*, 2016, **8**, 12337.

J. Lin, B. Hao, G. Cao, J. Wang, Y. Feng, X. Tan, and W. Wang, A study on the microbial community structure in oil reservoirs developed by water flooding, *J. Petrol. Sci. Eng.*, 2014, **122**, 354.

K.-L. Liu, K. Kondiparty, A. D. Nikolov, and D. Wasan, Dynamic Spreading of Nanofluids on Solids Part II: Modelling, *Langmuir*, 2012, **28**, 16274.

Q. Lv, Z. Li, B. Li, S. Li, and Q. Sun, Study of nanoparticle–surfactant-stabilized foam as a fracturing fluid, *Ind. Eng. Chem. Res.*, 2015, **54**, 9468.

R. N. Moghaddam, A. Bahramian, Z. Fakhroueian, A. Karimi, and S. Arya, Comparative study of using nanoparticles for enhanced oil recovery: wettability alteration of carbonate rocks, *Energy Fuels*, 2015, **29**, 2111.

T. Phenrat, H.-J. Kim, F. Fagerlund, T. Illangasekare, R. D. Tilton, and G. V. Lowry, Particle size distribution, concentration, and magnetic attraction affect transport of polymer-modified Fe_0 nanoparticles in sand columns, *Environ. Sci. Technol.*, 2009, **43**, 5079.

A. Pochert, S. Ziller, S. Kapetanovic, G. Neusser, C. Kranz, and M. Lindén, Intermediate pickering emulsion formation as a means for synthesizing hollow mesoporous silica nanoparticles, *New J. Chem.*, 2016, **40**, 4217.

W. Pu, C. Yuan, W. Hu, T. Tan, J. Hui, S. Zhao, S. Wang, and Y. Tang, Effects of interfacial tension and emulsification on displacement efficiency in dilute surfactant flooding, *RSC Adv.*, 2016, **6**, 50640.

F. Rao and Q. Liu, Froth treatment in Athabasca oil sands bitumen recovery process: A review, *Energy Fuels*, 2013, **27**, 7199.

A. Roustaei and H. Bagherzadeh, *J.* Experimental investigation of SiO_2 nanoparticles on enhanced oil recovery of carbonate reservoirs, *Petrol. Explor. Prod. Technol.*, 2015, **5**, 27.

Y. R. Saadat, N. Saeidi, S. Z. Vahed, A. Barzegari, and J. Barar, An update to DNA ladder assay for apoptosis detection, *BioImpacts*, 2015, **5**, 1.

E. Santini, E. Guzmán, M. Ferrari, and L. Liggieri, Emulsions stabilized by the interaction of silica nanoparticles and palmitic acid at the water–hexane interface, *Colloids Surf. A*, 2014, **460**, 333.

S. Sihler, A. Schrade, Z. Cao, and U. Ziener, Inverse Pickering emulsions with droplet sizes below 500 nm, *Langmuir*, 2015, **31**, 10392.

H. A. Son, S. K. Choi, E. S. Jeong, B. Kim, H. T. Kim, W. M. Sung, and J. W. Kim, Microbial activation of Bacillus subtilis-immobilized microgel particles for enhanced oil recovery, *Langmuir*, 2016, **32**, 8909.

B. Suleimanov, F. Ismailov, and E. Veliyev, Nanofluid for enhanced oil recovery, *J. Petrol. Sci. Eng.*, 2011, **78**, 431.

Q. Sun, Z. Li, S. Li, L. Jiang, J. Wang, and P. Wang, Utilization of surfactant-stabilized foam for enhanced oil recovery by adding nanoparticles, *Energy Fuels*, 2014, **28**, 2384.

D. J. Voorn, W. Ming, and A. M. van Herk, Polymer–Clay Nanocomposite Latex Particles by Inverse Pickering Emulsion Polymerization Stabilized with Hydrophobic Montmorillonite Platelets, *Macromolecules*, 2006, **39**, 2137.

K. Y. Yoon, H. A. Son, S. K. Choi, J. W. Kim, W. M. Sung, and H. T. Kim, Core flooding of complex nanoscale colloidal dispersions for enhanced oil recovery by in situ formation of stable oil-in-water Pickering emulsions, *Energy Fuels*, 2016, **30**, 2628.

H. Yu, Y. He, P. Li, S. Li, T. Zhang, E. Rodriguez-Pin, S. Du, S. Wang, S. Cheng, and C. W. Bielawski, Flow enhancement of water-based nanoparticle dispersion through microscale sedimentary rocks, *Sci. Rep.*, 2015, **5**, 8702.

J. Wang, G. Liu, L. Wang, C. Li, J. Xu, and D. Sun, Synergistic stabilization of emulsions by poly (oxypropylene) diamine and Laponite particles, *Colloids Surf. A*, 2010, **353**, 117.

A. J. Worthen, S. L. Bryant, C. Huh, and K. P. Johnston, Carbon dioxide-in-water foams stabilized with nanoparticles and surfactant acting in synergy, *AIChE J.*, 2013, **59**, 3490.

H. Zhang, A. Nikolov, and D. Wasan, Enhanced oil recovery (EOR) using nanoparticle dispersions: underlying mechanism and imbibition experiments, *Energy Fuels*, 2014, **28**, 3002.

D. Zhu, L. Wei, B. Wang, and Y. Feng, Aqueous hybrids of silica nanoparticles and hydrophobically associating hydrolyzed polyacrylamide used for EOR in high-temperature and high-salinity reservoirs, *Energies*, 2014, **7**, 3858.

Chapter 6: Emerging technologies and future outlook

The global demand for energy supply is increasing and few doubts remain that the world resources (renewable and fossil fuels) are adequately available. Because low efficiency and the high expense of renewable energy resources, currently their usage is not rational. And yet, for the world, the hydrocarbon resources remain as the main source of energy. However, most of the recoverable hydrocarbon resources have run out and using new technologies to recover ~70 % of remaining hydrocarbon is required. The best option, in reality, is using both conventional and unconventional resources including heavy and extra-heavy oil reserves, oil shale, tar sands, shale gas, tight gas, coal-bed methane, and natural-gas hydrates in the coming years (Al-Murayri et al., 2016). With the technical assistance of reservoir engineers, technologies will be increased and be performing within a few years (Figure 6.1).

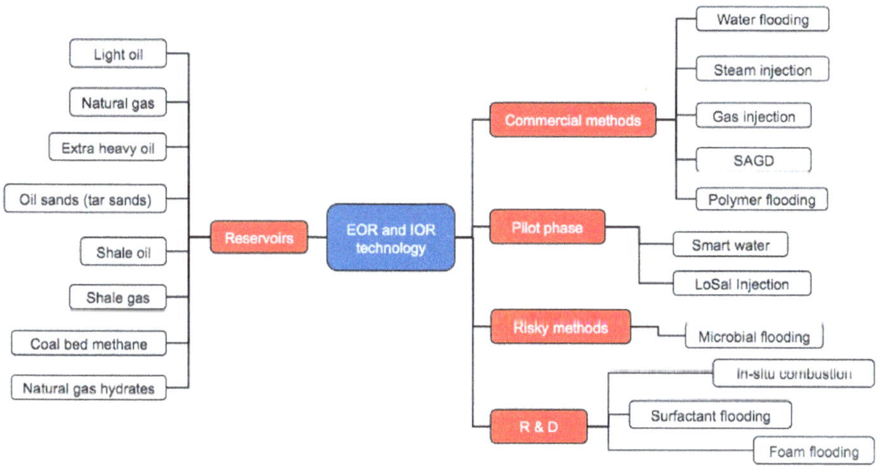

Figure 6.1: Summary of IOR/EOR technology, which can be utilized in real condition of the oil reservoirs. Adapted from T. Babadagli, Philosophy of EOR, *J. Pet. Sci. Eng.*, 2020, 188, 106930. Copyright: Elsevier (2020).

Currently, using low saline water, steam, miscible gas injection, polymer, and SAGD flooding process are being implemented in real reservoir environments. Given this, various technologies including smart water, acid gas injection, high-pressure air injection (HPAI), microbial, and foam must be developed over time (Gassara et al., 2015; Lin et al., 2014).

Medium and light oil reserves

Low salinity waterflooding

Ionic water content in the reservoir formation poses a problem (Myint & Firoozabadi, 2015). Low salinity water flooding, as a smart EOR flooding, is a promising technology designed to modify injected brine solution according to the reservoir's properties (Mahani et al., 2015; Nasralla et al., 2013; Austad et al., 2015). The main goal of this method is changing the wettability of rock from an oil-wet to a water-wet state. Because water flooding can result in various types of formation damage (e.g., fluids-rock incompatibility, organic and inorganic precipitation, and pore plugging), using smart water technology, an advanced water flooding, is a blueprint for the future.

According to data from the literature, in some cases, scaling issues arise in the reservoirs. The main reason for this phenomenon is the use of unsuitable EOR agents such as calcium carbonate when the temperature of oil formation rises, and the rate of salt precipitation increases. Typically, sodium chloride (NaCl), sulphate ($BaSO_4$, $SrSO_4$, $CaSO_4$), carbonate ($CaCO_3$, $FeCO_3$, $BaCO_3$, $SrCO_3$, $MgCO_3$, $ZnCO_3$, $PbCO_3$), and sulfide (PbS, ZnS, FeS_x) leads to scaling problems such as pore blocking and tubing flow restriction, especially in carbonate rocks. Basically, higher scaling tends to take place in the presence of heavy metals such as Ba^{2+}, SO_4^{2-}, BO_3^{3-}, Sr^{2+}, and PO_4^{3-}. On the other hand, adding lauric acid (Figure 6.2) and benzoic acid (Figure 6.3), to low saline water flooding increased fluid adsorption on surfaces due to the oil wetness characteristics as opposed to water-wet in oil reservoirs. This means that using different salts to decrease strong chemical adsorption on the rock surface is extremely problematic.

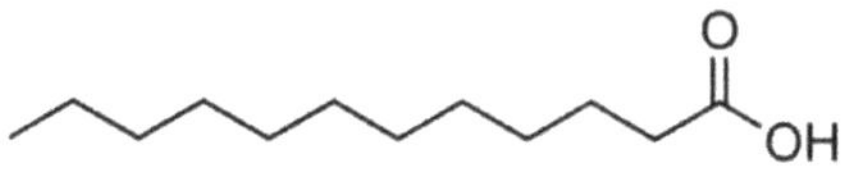

Figure 6.2: Structure lauric acid.

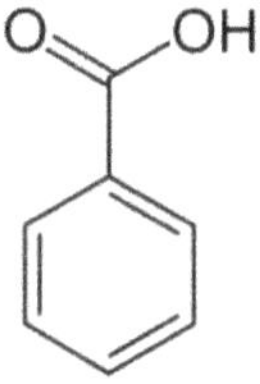

Figure 6.3: Structure of benzoic acid.

Modified seawater formulation significantly changes the adsorption of oil molecules on the rock surface by breaking down strong chemical reactions. Phosphate and polyphosphate compounds such as sodium hexametaphosphate (SHMP, Figure 6.4) are promising additives in smart water, which can reduce scaling problems at higher temperatures; however, the strong reaction of salts with water and rock surface generally occurs when sulfate being entered to the oil reserves. It was found that strong scaling tended to happen as a result of the precipitation and pore blocking upon mixing with water.

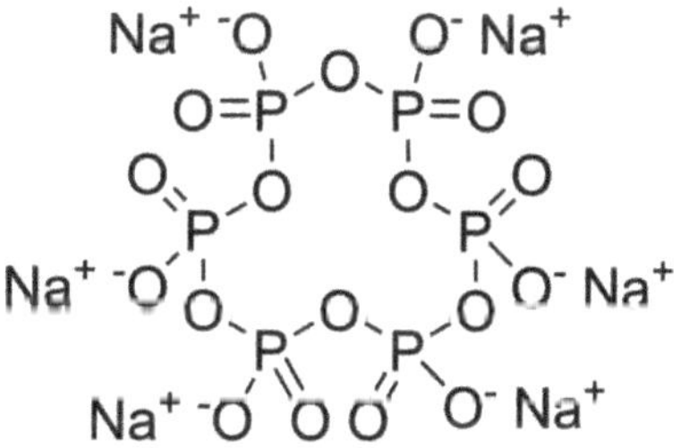

Figure 6.4: Structure of sodium hexametaphosphate (SHMP).

To prepare reservoirs for the next flooding with, for example, smart water such as smart water, engineers must optimize and assess the chemical interactions between the fluid-fluid and solid-fluid. Integrated methods, such as low salinity waterflooding, are economical and environmentally friendly methods, which can be applied, providing good geological properties are there (Figure 6.5).

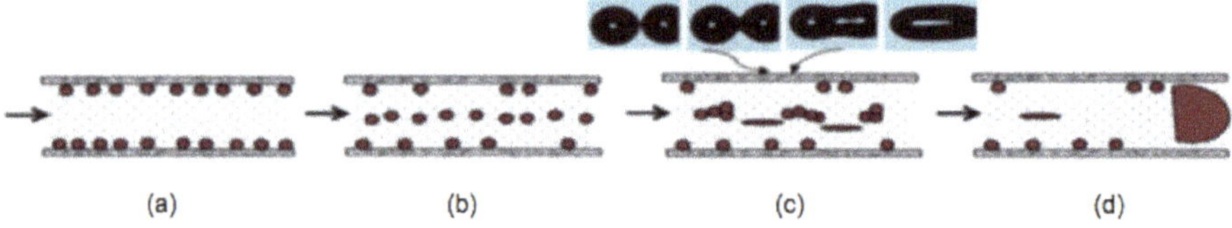

Figure 6.5: Schematic of the low salinity waterflooding EOR showing (a) smartwater injection, (b) oil drop release mainly due to rock fluid interfacial effects, (c) coalescence of oil drops due to fluid-fluid interfacial effects, and (d) oil bank due to combined effects from both interfaces. Reproduced with permission from Ayirala. S. S, Saleh. S. H, Yousef. A. A, Microscopic scale study of individual water ion interactions at complex crude oil-water interface: a new smartwater flood recovery mechanism, *SPE Improved Oil Recovery Conference*, 2016. Copyright: SPE (2016).

Researchers, based on laboratory experiments, have realized the importance of multivalent cations (such as calcium and magnesium) for successful smart water flooding (Sohal et al., 2016). They are now focusing on manipulating brines before each flooding to form a positive wettability alteration with lower adsorption on the rock surface. Prior to applying any smart water flooding, reducing chemical interactions between fluid-fluid and solid-fluid interfaces is essential. Therefore, finding optimal ionic strength can be a catalyst to decrease the strong adsorption between oil and rock surface. The different reactions of mono/divalent cations in water and on the carbonated rock (Chalk) were carried out using spontaneous imbibition experiments (Zhang et al., 2007), while complex interactions between oil-water formed, the addition of a defined amount of water needed to decrease bonding. Core flooding experiments on non-wet rock revealed a 3-5% surge in original oil in place (OOIP) and 16-21% after water and LoSal water flooding, respectively. Typically, low salinity water flooding can be used in both carbonated and sandstone reservoirs such as the North Sea and Middle East oilfields (Nasralla et al., 2013). High oil recovery (26%) was obtained after flooding optimal saline water solution in the Endicott oilfield, Alaska by BP. In 2016, BP,

ConocoPhillips, Chevron, and Shell will initiate the largest offshore low salinity water-flooding project Clair Ridge oilfield (in the North Sea). Clair Ridge is expected to produce more than 40 million barrels of additional oil at relatively low cost.

Would seawater flooding work instead?

The short answer is no. Seawater is a solution containing a large amount of divalent and monovalent cation and anions. It has already been confirmed that using seawater increases the binding energy between salt/oil/rock. To use seawater safely, engineers need desalination equipment (Cohen et al., 2012). A desalination unit performs as a membrane that can tune water chemistry. High oil recovery can be achieved if the optimal low salinity water flooding is there. Polymers and other EOR additives can also be added to low brine solution to decrease the cost of EOR operations by recycling and desalting produced water. Therefore, a sharp increase in oil recovery (5-7%) is feasible by using the in-site desalination unit on seawater solution (under 5000 ppm brine).

Companies are working to build a highly efficient desalination unit using new technologies. Forward/reverse osmosis and membranes are the main methods proposed to improve desalination performance. On the other hand, some nanomaterials such as carbon nanotubes, graphene oxide, and metal nanoparticles (NPs) can potentially be used.

The mechanism of low salinity water flooding

LoSal water flooding mechanism is still an unanswered question. However, researchers essentially explain this as changing rock wetting properties, which can be described by two theories: (a) multicomponent ion exchange (MIE), and (b) expansion of electrical double layer (EDL).[1] The reservoir pH can be fine-tuned by flooding optimal low salinity. Typically, ionic exchange on the rock matrix sites is described by MIE theory. According to MIE theory, unfavorable cation adsorption takes place in the presence of Mg^{2+} and Ca^{2+} on

---

[1] An electrical double layer, or simply double layer (DL), is a structure that appears on the surface of an object when it is exposed to a fluid.

the rock and decreases the oil liberation. According to MIE theory, polar organic and organometallic complexes can be removed from the surfaces and be replaced with other cation charges such as Na^+. The main reason for oil liberation is the presence of weak van der Waals interactions on a surface. In clay-based formations, cationic exchange mostly takes place via quaternary ammonium compounds or heterocyclic ring with metal cations on the clay. Low salinity water flooding can remove the Ca^{2+} and Mg^{2+}; decrease residual oil saturation (5.6%-7.6%) in Clair's oilfield. Microscopic displacement efficiency is improved by eliminating strong chemical bindings between brine-oil-rock, resulting in alterations in rock wettability.

The most important reactions between rock surface-fluid in different scenarios are shown in Table 6.1. In LoSal injection, cation/ligand exchange and cation/water bridging will be greatly affected during low salinity.

Mechanism	Organic functional group involved
Cation exchange	Amino, ring NH, heterocyclic N (aromatic ring)
Protonation	Amino, heterocyclic N, carbonyl, carboxylate
Anion exchange	Carboxylate
Water bridging	Amino, carboxylate, carbonyl, alcoholic OH
Cation bridging	Amines, carboxylate, carbonyl, alcoholic OH
Ligand exchange	Carboxylate
Hydrogen bonding	Amino, carboxylate, carbonyl, phenolic OH
van der Waal interactions	Uncharged organic units

Table 6.1: Mechanism of association organic functional groups-soil minerals

Alkaline flooding

Alkaline flooding (caustic flooding) is an EOR method similar to low salinity water flooding. However, in this process only an alkaline solution such as $NH_3.H_2O$, NH_4OH, Na_2CO_3 and $NaOH$ can be used. Forming surfactant consists of alkaline interaction with oil. Therefore, the idea behind surfactant is decreasing IFT between oil and water that is an ideal scenario for oil recovery (Pei et al., 2012). Flooding alkaline solution in carbonated reservoirs is not recommended because it produces hydroxide precipitation and leads to the attachment of caustics, scaling, and finally pore plugging in carbonate rocks. This situation in the reservoir will occur in porous media when calcium and magnesium are in abundance.

NPs flooding

Research on the use of new technologies in oil recovery has continued, for a decade in which researchers focused on applying NPs as agents inside the pores. The addition of NPs to rock alters the fluid-fluid and rock-fluid interfaces and improves wetting properties of rock to increase oil sweeping. NPs flooding might also be combined with surfactant and polymer flooding to increase oil relative permeability (rel-perm) and decrease water rel-perm. In this approach, water-cut would be decreased inside the pores and pore-throats. Moreover, it may be possible to employ as a catalyst at a higher temperature in highly viscous reserves using In-situ oil recovery approach (Hamedi-Shokrlu & Babadagli, 2014). So that they make a reservoir an underground refinery, which minimizes viscosities of heavy and extra-heavy oil e.g., shale or bitumen. W, Ni, and Mo oxides are possible choices to decrease viscosities of unconventional oil resources. TiO_2 nanofluid (0.01 wt. %) improve the rock wettability in the presence of heavy oil and lead to an increase of 49% in oil recovery after water-flooding. Researchers have explored adding –OH groups (on the NPs) which caused a decrease in chemical interactions between oil-rock and decrease NPs tendency to absorb on the rock substrate.

NPs flooding obstacles

Traditional EOR methods (water, gas, steam, and polymer flooding) have been in use for a long time, but significant technical, operational and economic factors continue to limit their application. Several completely new technologies such as highly stable polymers, surfactants, and NPs have been developed to improve oil recovery. These additives have shown great effectiveness on EOR parameters, but they are, at present, not cost-effective in comparison to current EOR techniques. Developing the facile injection scenario based on NPs is an important approach for oil companies.

A low-cost preparation method, eco-friendly precursor, mass-production scalable, and high stability based on the reservoir mediums are still mandatory to develop EOR procedures. Effective use of NPs across a wide range of EOR applications, rely on particle stability through oil-pathways in inter-connected pores. Therefore, considerable research efforts have built stabilized NPs in the reservoir. Fine migration represents the best example of the issue. When the fine particles are highly precipitated in the reservoir, their interactions result in plugging of the pores and reducing rock permeability.

Heavy and extra heavy oil reserves

In-situ combustion

In-situ combustion (ISC) is a mixture of fire with air or O_2 injection to generate heat in-place and higher temperatures to decrease fluid viscosity into the reservoir. In-situ combustion is divided into two subgroups: wet and dry combustion. If the water added to air makes an oil front zone in the vicinity of the wellbore, it is called *wet combustion*. On the other hand, if only the air is utilized for ignition it is called *dry combustion*. Understanding the continuous air injection mechanism aids the creation of a burning front zone and complete liquid sweeping toward the wellbore. Non-thermal methods are a potential route for light and moderate oil reservoirs (100-2000 cp) these methods are not appropriate for thermal injection. However, scientists have claimed using in-situ combustion leads to corrosion and generates toxic gasses in the reserves.

Researchers classified the trapped oil in reservoirs based on their viscosity. Light oil (API gravity >22), the heavy (API gravity 10-20), and extra heavy (API gravity <10) are too viscous and do not have the mobility to flow through the porous media. The reason that oil is classified as a heavy and extra heavy is due to a considerable amount of sulfur, nitrogen, oxygen, and high C/H unit ratio (Alvarado & Manrique, 2010).

Bitumen recovery

Many efforts have been made to decrease the number of aromatics, resins, and asphaltenes in the reserves. Some applicable technologies such as cycle steam simulator (CSS), steam flooding, steam-assisted gravity drainage (SAGD), and in-situ combustion (ISC) work by breaking interactions between highly polar components, oil, and rock with decreasing oil viscosity. Bitumen extraction methods such as injecting steam and heat are effective in reducing viscosity. Researchers also have explored techniques using VAPEX, ES-SAGD, and SAGP to decrease heavy oil viscosity, but their applications thus far are not cost-effective (Flury et al., 2014). The presence of solvent gas (ethane, propane, and butane) with a carrier gas (N_2 or CO_2) instead of steam is called VAPEX. Combining low steam value (10%) with the solvent gas leads to a solvent expansion in the VAPEX; therefore, it is called expanding solvent SAGD (ES-SAGD). CH_4 or N_2 is injected with steam to push gas into SAGD, is called steam and gas push (SAGP). The N_2 injection can be considered as a sweeping agent in the high-pressure reservoir, e.g., the Cantarell field in Mexico as the largest N_2 injection in the world.

Moreover, ionic liquids (ILs) and organic solvents can be used to increase the amount of bitumen extraction in tar sands. Particle size, shape, surface area, sorption properties, and active surface sites are the main parameters in oil upgrading. Ultra-dispersed nanosized catalysts can be injected into the reservoir at minimal heat requirement owing to easily bitumen transportation via pipelines. Moreover, the effect of cerium oxide nanoparticles (CeO) has been investigated by catalytic cracking reaction and supercritical water in Canadian oil sand bitumen (Figure 6.5).

Desulfurization and emulsification

Desulfurization and emulsification can enhance the viscosity of heavy oil. Oil-oxidative desulfurization (ODS) method as a demulsification method can decrease the polarity of heavy oil components and upgrade oil for oil sands derived bitumen. ODS has been developed with

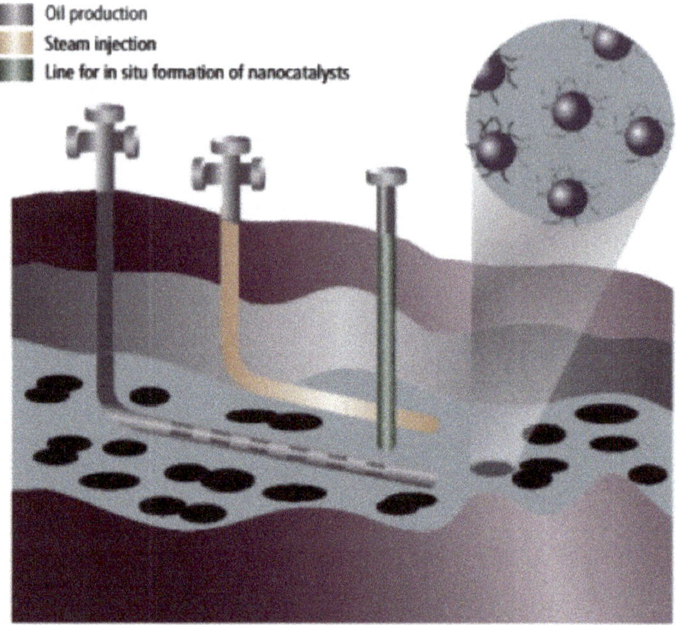

Figure 6.5: Schematic representation of in-situ upgrading with ultra-dispersed nanoparticles in sand oil. Reproduced with permission from R. Hashemi, N. N. Nassar, P. P. Almao, Enhanced heavy oil recovery by in situ prepared ultradispersed multimetallic nanoparticles: A study of hot fluid flooding for Athabasca bitumen recovery, *Applied Energy,* **2014, 133, 374. Copyright: Elsevier (2013).**

oxidization of sulfur using Canadian Cold Lake bitumen. An emulsification method was deployed in the presence of polyvinyl alcohol (Figure 6.6) and evaluated the effect of surfactant molecular weight and degree of hydrolysis for reducing the viscosity of Canadian heavy oil.

Figure 6.6: Structure of polyvinyl alcohol (PVA).

Tar sand

Over previous decades, progress in the extraction of tar sand (Figure 6.7) has emerged as the fastest-growing area around the world. Mostly, oil sand consists of sand-clay (86%), and bitumen-water (14%). In situ mining and surface, mining has been applied successfully to recover trapped oil in the oil sand. Currently, cycle steam simulator (CSS-steam), steam-assisted gravity drainage (SAGD-hot steam), and supercritical water (SCW) are methods that can recover bitumen from deep underground locations. However, specialist equipment should be used for the separation of approximately 90% bitumen from the rock. It exhibited many advantages for converting (upgrading) bitumen or heavy oil to lighter fractions (synthetic crude oil). It should be noted that adding hydrogen to higher components (hydrocracking) or carbon removing (coking) simplifies the process for obtaining valuable light fractions of oil. The importance of environmental footprints, CO_2 emissions, wastewater, and oil sands infrastructure should be considered.

Figure 6.7: Oil sands on the banks of the Athabasca River (c. 1900 - 1930). Copyright: Library and Archives Canada.

Shale oil

Shale oil is a class of mature oil. The majority of shale oil is kerogen (a solid polymer) and organic bitumen, which are located in low permeability formation at shallow depths. The conversion of kerogen to oil requires high temperatures in the absence of oxygen.

Hydraulic fracturing

Hydraulic fracturing (or fracking) is the main method to boost shale oil production in low/ultralow permeability pores (ranging from 10 to hundred nanometers). Hydraulic fracturing consists of pumping a fluid into wells in order to increase pressure and produce fractures within the rock formation. To keep the fracture open after the injection stops, sand with high permeability is added to the fracture. The schematic of fluid transport into the shale reservoir is shown in Figure 6.8.

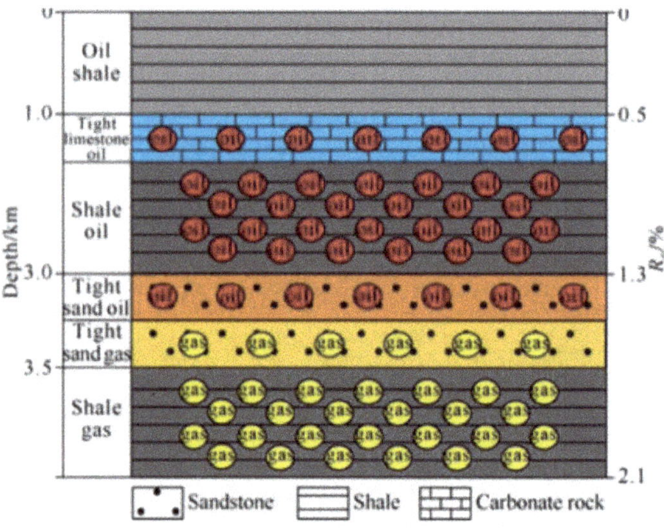

Figure 6.8: Hydrocarbon accumulation patterns in shale measures. Reproduced with permission from Caineng. Z, Zhi. Y, Jingwei. C, Rukai. Z, Lianhua. H, Shizhen. T, Xuanjun. Y, Songtao. W, Senhu. L, Lan. W, Formation mechanism, geological characteristics and development strategy of nonmarine shale oil in China, *Petrol. Explor. Dev.*, 2013, 40, 15.

Functionalized aluminate such as polymers/resin proppants, resin-coated sand, sintered bauxite and gel compositions are highly effective chemicals in keeping fractures open and suspend the functionalized aluminate structure (Gomez et al., 2017; Akhtarmanesh et al., 2013). To combat high interfacial tension caused by adding aluminate components, linear/crosslinked guar gum (Figure 6.9), viscoelastic surfactants, and energized fluids (a large fraction of gas) can be used to reduce IFT and increase the fluid transport. "slickwater" is another approach by which adding a low concentration of polyacrylamide (Figure 6.10) and linear gel (reducer) is carried out to control the conductivity damage through the fractures. The main focus of the materials used in hydraulic fracturing is developing a cost-effective and HPHT stable chemicals similar to guar gum and polyacrylamides.

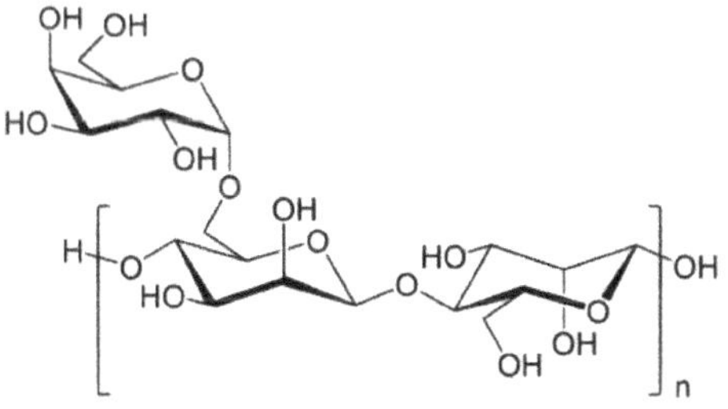

Figure 6.9: Structure of a guaran unit of guar gum.

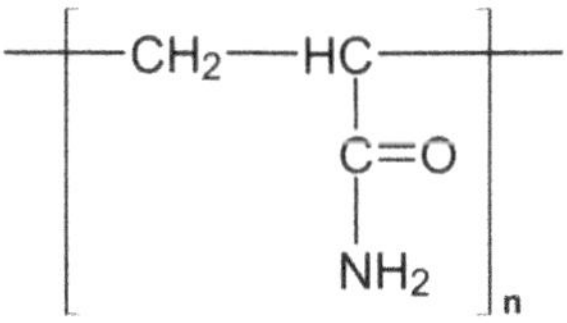

Figure 6.10: Structure of polyacrylamide.

Adding nanoparticle to shale reservoirs improve the viscoelasticity of injected fluid during a water loss. Recently, it was suggested that the silica nanofluids can be utilized as an agent in low permeability rocks (shale oil), where the surface of silica NPs were modified with vinyl-triethoxysilane (Figure 6.11) and 2-mercaptobenzimidazole (Figure 6.12). It showed good stability in different pH and NaCl concentrations, proper interface tension (IFT) and wettability alteration toward a water-wet state.

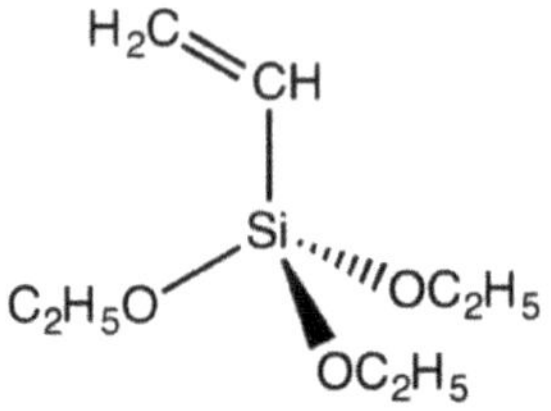

Figure 6.11: Structure of vinyltriethoxysilane.

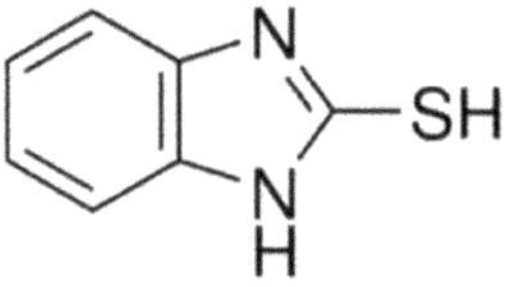

Figure 6.12: Structure of 2-mercaptobenzimidazole.

Oil shale pyrolysis using a solid heat carrier (SHC) method is a useful technique in changing shale oil viscosity in a circulating fluidized bed (CFB), where shale is heated inside the well. Based on this system, the oil will be boiled and vaporized through the shale reservoirs and consequently, extracted as light fractions of the surface as shown in Figure 6.13. Researchers revealed using SHC led to the formation of lighter oil fractions via electrical heaters in vertically drilled wells. Inducing heat in the shale reservoir transformed lighter fractions of shale and high oil recovery after 2-3 years.

Tight gas

The production of natural gas from low permeability mediums (below 0.1 mD) in tight rock is called tight gas. Moreover, gas production from shale formation is called shale gas. In recent years, gas production from the tight reservoirs has drastically increased. Vertical wells are not suitable for gas extraction because they are unsuitable in low permeable reservoirs. For this reason, the horizontal wells are a good choice to increase the amount of gas recovery. Production of tight and shale gas requires hydraulic fracturing or horizontal wells. Horizontal wells provide higher contact with deposit shale compared to vertical wells and enables the effective gas transfer. Today's technology is only suitable for onshore production and offers a maximum recovery rate of 20% of the volume in place.

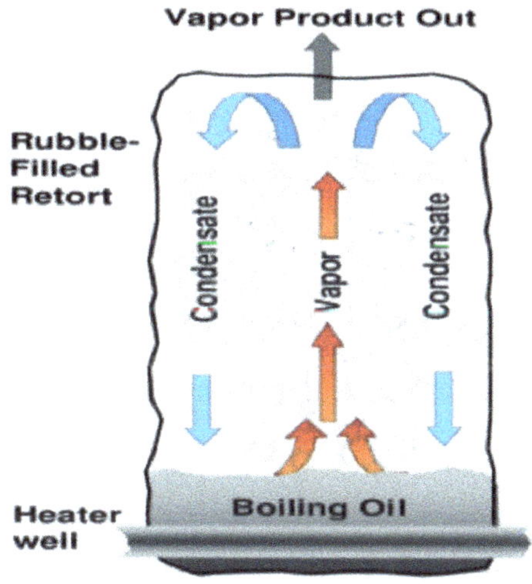

Figure 6.13: Schematic representation of refluxing oil in oil shale formation. Reproduced with permission from Burnham. A. K, Day. R. L, Hardy. M. P, Wallman. P. H, *Oil Shale: A Solution to the Liquid Fuel Dilemma*, **Ed. O. I. Ogunsola, A. M. Hartstein, and O. Ogunsola, Oxford University Press: Oxford (2010).**

In coal deposits, significant amounts of methane-rich gas are generated and stored within the coal structure. The gas is normally released during mining, although more recent research aimed to capture more gas. This process is applied not only for economic exploitation, but also for safety and environmental reasons. Coal-bed methane (CBM), however, is typically methane gas trapped within coal deposits that are not profitable for extraction due to high depths or poor coal quality. Coal beds have low permeability that reduces with increasing depth. Therefore, hydraulic fracturing and/ or horizontal wells are required ease the fluid to flow through a well. Because of the pressure, water permeates into coal and traps the gas. It is then extracted again thus reducing the pressure and enabling methane to flow out of the coal through the well. Figure 6.14 shows a typical production curve of CBM, with volumes of methane and water production over time.

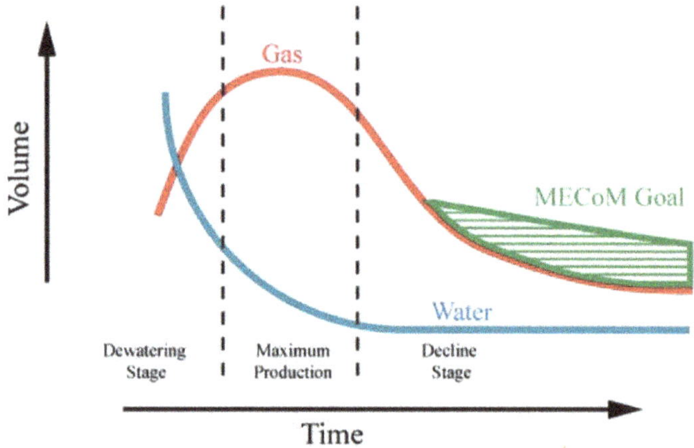

Figure 6.14: Production curve of CBM, with volumes of methane and water production over time. Reproduced with permission from D. Ritter, D. Vinson, E. Barnhart, D. M. Akob, M. W. Fields, A. B. Cunningham, W. Orem, and J. C. McIntosh, Enhanced microbial coal bed methane generation: a review of research, commercial activity, and remaining challenges, *Int. J. Coal Geo.*, 2015, 146, 28. Copyright: Elsevier (2015).

In the first phase, a large amount of contaminated water is produced, which is usually re-injected into the formations. Today's research efforts aim to develop techniques based on CO_2 injections into coal bed formations to enhance methane production. The easy CO_2 adsorption of coal helps release methane and offers significant potential for CO_2 geological storage and reduction of CO_2 in the atmosphere.

The swelling of smectite clays in the inter burden rock layers of coal seam gas wells results in spalling of fine particles that may negatively impact gas production through damage to the well's pump and/or the permeability of the coal layers. One cause of the clay swelling is the change in water chemistry at the wellbore location due to drilling fluids and the influx of produced water. The method used to mitigate the swelling of smectite clays in oil and gas reservoirs is to stabilize the clay using a brine such as 4% KCl, although this technology is relatively low cost and initially effective, the mitigation of clay swelling is temporary because K^+ ions are easily washed off the clay when the well is brought into production. An alternative approach recently reported is the use of nanoparticles and nanofluids to control clay swelling.

Five commercially available NPs (SiO_2, Al_2O_3, ZnO, Fe_2O_3, and ZrO_2) have been evaluated for the prevention of swelling of natural bentonite clay, rich in sodium montmorillonite, $[(Na,Ca)_{0.33}(Al,Mg)_2(Si_4O_{10})(OH)_2 \cdot nH_2O]$. The effectiveness of the nanoparticles to prevent clay swelling was measured with a visual swelling index method based on ASTM D5890-11 and compared to the swelling of the clay in 4% KCl brine. In the initial nanoparticle screening tests performed in distilled water, all the NPs except for the ZnO exhibited some potential to mitigate the swelling of bentonite. The next stage of screening experiments was performed in model formation water solutions containing 2500 mg /L and 9000 mg/L of Na ions at pHs of 5 and 9. In the model formation water tests, the SiO_2 was the most effective nanoparticle to mitigate clay swelling across the range of conditions examined. These results suggest NPs may be a potential solution to mitigate clay swelling and spalling in coal seam gas reservoirs as well as other types of reservoirs. Further research is required to elucidate the mechanism of swelling inhibition with SiO_2 NPs and to develop practical methods to deploy NPs into a coal seam gas (CSG) well.

Natural-gas hydrates

Natural-gas hydrates (also known as methane clathrates[1]) are crystalline materials in which gas molecules are surrounded by a lattice of water molecules. They are formed by water and natural gas (methane)

---

[1] A clathrate is a chemical substance consisting of a lattice that traps or contains molecules including host–guest complexes and inclusion compounds.

at high pressures and low temperatures (Figure 6.15). In such conditions, they are stable and only dissociate very slowly. At present, within the oil and gas industry, natural gas hydrates are seen as a problem rather than as a resource. Formation of "snow-like" hydrates can damage oil and gas pipelines and cause problems in drilling pipes; however, the vast potential of gas trapped within hydrates offers a large resource.

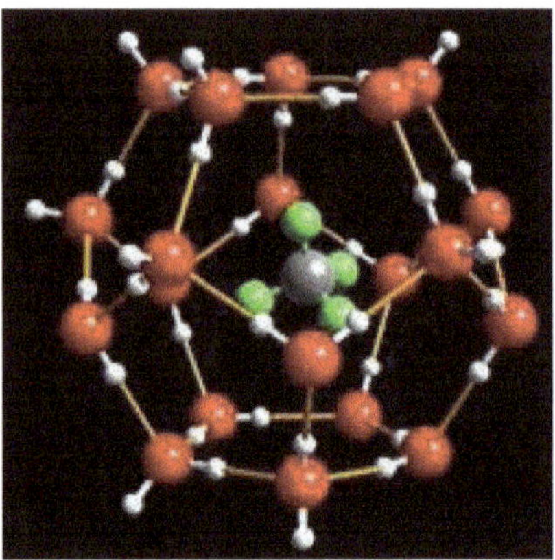

Figure 6.15: The molecular level structure of the smallest gas hydrate consisting of methane molecules (CH_4, green and gray) embedded in lattices of water molecules (H_2O, red and white).

Three basic methods exist for gas hydrate exploitation as an energy resource: depressurization, thermal injection and inhibitor injection. In some cases, hydrates are located above gas reservoirs and dissociate as the production from natural gas reduces the underground pressure. Depressurization is, therefore, the easiest method to extract hydrates but does come with a few technical challenges. However, well depressurization does not necessarily reduce the pressure of the entire hydrate layer. Thermal injection techniques involve the injection of steam or hot water into the well to decompose the hydrates and thus generate the gas. One challenge in this process is that hydrates are often found at deep locations and injected fluids are cooled before reaching the hydrate layer. Inhibitor injection techniques are used to collapse the crystalline hydrate in offshore natural-gas pipelines to prevent hydrate formation. Inhibitors such as methanol dissolve methane from the hydrate, and the gas is released.

The average hydrate composition is 1 mole of CH_4 for every 5.75 moles of H_2O (i.e., $4CH_4 \cdot 23H_2O$), though this is dependent on how many methane molecules *fit* into the various cage structures of the water lattice. The observed density is around 0.9 g/cm^3. One volume of methane hydrate solid would, therefore, contain, on average, 168 volumes of CH_4 gas.

Bibliography

M. Abdulbaki, C. Huh, K. Sepehrnoori, M. Delshad, and A. Varavei. A critical review on use of polymer microgels for conformance control purposes, *J. Petrol. Sci. Eng.*, 2014, **122**, 741.

S. S. Ayirala, S. H. Saleh, and A. A. Yousef, Microscopic scale study of individual water ion interactions at complex crude oil-water interface: a new smartwater flood recovery mechanism, *SPE Improved Oil Recovery Conference*, 2016.

S. Akhtarmanesh, M. A. Shahrabi, and A. Atashnezhad, Improvement of wellbore stability in shale using nanoparticles, *J. Petrol. Sci. Eng.*, 2013, **112**, 290.

P. P. Almao, In situ upgrading of bitumen and heavy oils via nanocatalysis, *Can. J. Chem. Eng.*, 2012, **90**, 320.

M. T. Al-Murayri, B. B. Maini, T. G. Harding, and J. Oskouei, Multicomponent solvent co-injection with steam in heavy and extra-heavy oil reservoirs, *Energy Fuels*, 2016, **30**, 2604.

V. Alvarado and E. Manrique, Enhanced oil recovery: an update review, *Energies*, 2010, **3**, 1529.

T. Austad, S. F. Shariatpanahi, S. Strand, H. Aksulu, and T. Puntervold, Low salinity EOR effects in limestone reservoir cores containing anhydrite: a discussion of the chemical mechanism, *Energy Fuels*, 2015, **29**, 6903.

T. Babadagli, Philosophy of EOR, *J. Pet. Sci. Eng.*, 2020, **188**, 106930.

A. K. Burnham, R. L. Day, M. P. Hardy, and P. H. Wallman, *Oil Shale: A Solution to the Liquid Fuel Dilemma*, Ed. O. I. Ogunsola, A. M. Hartstein, and O. Ogunsola, Oxford University Press: Oxford (2010).

Z. Caineng, Y. Zhi, C. Jingwei, Z. Rukai, H. Lianhua, T. Shizhen, Y. Xuanjun, W. Songtao, L. Senhu, and W. Lan, Formation mechanism, geological characteristics and development strategy of nonmarine shale oil in China, *Petrol. Explor. Dev.*, 2013, **40**, 15.

G. Cheraghian, M. Hemmati, M. Masihi, and S. Bazgir, An experimental investigation of the enhanced oil recovery and improved performance of drilling fluids using titanium dioxide and fumed silica nanoparticles, *J. Nanostructure Chem.*, 2013, **3**, 78.

D. Cohen-Tanugi and J. C. Grossman, Water desalination across nanoporous graphene, *Nano letters*, 2012, **12**, 3602.

H. Ehtesabi, M. M. Ahadian, V. Taghikhani, and M. H. Ghazanfari, Enhanced heavy oil recovery using TiO_2 nanoparticles: investigation of deposition during transport in core plug, *Energy Fuels*, 2013, **28**, 423.

C. Flury, A. Afacan, M. Tamiz Bakhtiari, J. Sjoblom, and Z. Xu, Effect of caustic type on bitumen extraction from Canadian oil sands, *Energy Fuels*, 2013, **28**, 431.

V. Gomez, S. Alexander, and A. R. Barron, Proppant immobilization facilitated by carbon nanotube mediated microwave treatment of polymer-proppant structures, *Colloids Surf. A*, 2017, **513**, 297.

M. Jafarnezhad, M. S. Giri, and M. Alizadeh, Impact of SnO_2 nanoparticles on enhanced oil recovery from carbonate media, *Energ. Source. Part A*, 2017, **39**, 121.

R. Hashemi, N. N. Nassar, and P. Pereira-Almao, Enhanced heavy oil recovery by in situ prepared ultradispersed multimetallic nanoparticles: A study of hot fluid flooding for Athabasca bitumen recovery, *Energy Fuels*, 2013, **27**, 2194.

H. Mahani, A. L. Keya, S. Berg, W.-B. Bartels, R. Nasralla, and W. R. Rossen, Insights into the mechanism of wettability alteration by low-salinity flooding (LSF) in carbonates, *Energy Fuels*, 2015, **29**, 1352.

H. Pei, G. Zhang, J. Ge, M. Tang, and Y. Zheng, Comparative effectiveness of alkaline flooding and alkaline–surfactant flooding for improved heavy-oil recovery, *Energy Fuels*, 2012, **26**, 2911.

D. Ritter, D. Vinson, E. Barnhart, D. M. Akob, M. W. Fields, A. B. Cunningham, W. Orem, and J. C. McIntosh, Enhanced microbial coalbed methane generation: a review of research, commercial activity, and remaining challenges, *Int. J. Coal Geo.*, 2015, **146**, 28.

Shariatpanahi. S. F, Hopkins. P, Aksulu. H, Strand. S, Puntervold. T, Austad. T, Water based EOR by wettability alteration in dolomite, *Energy Fuels*, 2016, **30**, 180.

J. Sheng, *Enhanced oil recovery field case studies*, Gulf Professional Publishing (2013).

M. A. Sohal, G. Thyne, and E. G. Søgaard, Review of recovery mechanisms of ionically modified waterflood in carbonate reservoirs, *Energy Fuels*, 2016, **30**, 1904.

S. Trabelsi, J.-F. O. Argillier, C. Dalmazzone, A. Hutin, B. Bazin, and D. Langevin, Effect of added surfactants in an enhanced alkaline/heavy oil system, *Energy Fuels*, 2011, **25**, 1681.

J. Wang, G. Liu, L. Wang, C. Li, J. Xu, and D. Sun, Synergistic stabilization of emulsions by poly(oxypropylene) diamine and laponite particles, *Colloids Surf. A*, 2010, **353**, 117

Zhang. Z, Yang. X, Jia. H, Zhang. H, Kerogen beneficiation from Longkou oil shale using gravity separation method, *Energy Fuels*, 2016, **30**, 2841.